Separation, Recovery, and Purification in Biotechnology

ACS SYMPOSIUM SERIES **314**

Separation, Recovery, and Purification in Biotechnology

Recent Advances and Mathematical Modeling

Juan A. Asenjo, EDITOR
Columbia University

Juan Hong, EDITOR
Illinois Institute of Technology

Developed from a symposium sponsored by
the Division of Microbial and Biochemical Technology
at the 190th Meeting
of the American Chemical Society,
Chicago, Illinois,
September 8–13, 1985

American Chemical Society, Washington, DC 1986

Library of Congress Cataloging-in-Publication Data

Separation, recovery, and purification in biotechnology.
  (ACS symposium series, ISSN 0097-6156; 314)
  Includes bibliographies and index.
  1. Biotechnology—Technique—Congresses.
  2. Biomolecules—Purification—Congresses.
  3. Biological chemistry—Technique—Congresses.
    I. Asenjo, Juan A., 1949-   . II. Hong, Juan.
  III. American Chemical Society. Meeting (190th: 1985: Chicago, Ill.) IV. Series.
  TP248.24.S47   1986   660′.6′028   86-10833
  ISBN 0-8412-0978-2

Copyright © 1986

American Chemical Society

All Rights Reserved. The appearance of the code at the bottom of the first page of each chapter in this volume indicates the copyright owner's consent that reprographic copies of the chapter may be made for personal or internal use or for the personal or internal use of specific clients. This consent is given on the condition, however, that the copier pay the stated per copy fee through the Copyright Clearance Center, Inc., 27 Congress Street, Salem, MA 01970, for copying beyond that permitted by Sections 107 or 108 of the U.S. Copyright Law. This consent does not extend to copying or transmission by any means—graphic or electronic—for any other purpose, such as for general distribution, for advertising or promotional purposes, for creating a new collective work, for resale, or for information storage and retrieval systems. The copying fee for each chapter is indicated in the code at the bottom of the first page of the chapter.

The citation of trade names and/or names of manufacturers in this publication is not to be construed as an endorsement or as approval by ACS of the commercial products or services referenced herein; nor should the mere reference herein to any drawing, specification, chemical process, or other data be regarded as a license or as a conveyance of any right or permission, to the holder, reader, or any other person or corporation, to manufacture, reproduce, use, or sell any patented invention or copyrighted work that may in any way be related thereto. Registered names, trademarks, etc., used in this publication, even without specific indication thereof, are not to be considered unprotected by law.

PRINTED IN THE UNITED STATES OF AMERICA

ACS Symposium Series

M. Joan Comstock, *Series Editor*

Advisory Board

Harvey W. Blanch
University of California—Berkeley

Alan Elzerman
Clemson University

John W. Finley
Nabisco Brands, Inc.

Marye Anne Fox
The University of Texas—Austin

Martin L. Gorbaty
Exxon Research and Engineering Co.

Roland F. Hirsch
U.S. Department of Energy

Rudolph J. Marcus
Consultant, Computers &
 Chemistry Research

Vincent D. McGinniss
Battelle Columbus Laboratories

Donald E. Moreland
USDA, Agricultural Research Service

W. H. Norton
J. T. Baker Chemical Company

James C. Randall
Exxon Chemical Company

W. D. Shults
Oak Ridge National Laboratory

Geoffrey K. Smith
Rohm & Haas Co.

Charles S. Tuesday
General Motors Research Laboratory

Douglas B. Walters
National Institute of
 Environmental Health

C. Grant Willson
IBM Research Department

FOREWORD

The ACS SYMPOSIUM SERIES was founded in 1974 to provide a medium for publishing symposia quickly in book form. The format of the Series parallels that of the continuing ADVANCES IN CHEMISTRY SERIES except that, in order to save time, the papers are not typeset but are reproduced as they are submitted by the authors in camera-ready form. Papers are reviewed under the supervision of the Editors with the assistance of the Series Advisory Board and are selected to maintain the integrity of the symposia; however, verbatim reproductions of previously published papers are not accepted. Both reviews and reports of research are acceptable, because symposia may embrace both types of presentation.

CONTENTS

Preface ix

PRODUCT RELEASE AND RECOVERY

1. Protein Release from Chemically Permeabilized *Escherichia coli* 2
   David J. Hettwer and Henry Y. Wang

2. Structured and Simple Models of Enzymatic Lysis and Disruption of Yeast Cells 9
   J. B. Hunter and J. A. Asenjo

3. Dual Hollow-Fiber Bioreactor for Aerobic Whole-Cell Immobilization 32
   Ho Nam Chang, Bong Hyun Chung, and In Ho Kim

4. A Membrane Reactor for Simultaneous Production of Anaerobic Single-Cell Protein and Methane 43
   R. K. Finn and E. Ercoli

SEPARATION AND CONCENTRATION PROCESSES

5. Membrane Processes in the Separation, Purification, and Concentration of Bioactive Compounds from Fermentation Broths 52
   Enrico Drioli

6. Liquid Emulsion Membranes and Their Applications in Biochemical Separations 67
   M. P. Thien, T. A. Hatton, and D. I. C. Wang

7. Use of Aqueous Two-Phase Systems for Recovery and Purification in Biotechnology 78
   Bo Mattiasson and Rajni Kaul

8. Recovery of Proteins from Polyethylene Glycol–Water Solution by Salt Partition 93
   G. B. Dove and G. Mitra

9. Modeling of Precipitation Phenomena in Protein Recovery 109
   C. E. Glatz and R. R. Fisher

PURIFICATION OPERATIONS

10. Process Considerations for Scale-Up of Liquid Chromatography and Electrophoresis 122
    S. R. Rudge and M. R. Ladisch

11. Mathematical Modeling of Bioproduct Adsorption Using Immobilized Affinity Adsorbents 153
    Somesh C. Nigam and Henry Y. Wang

12. High-Resolution, High-Yield Continuous-Flow Electrophoresis 169
    William A. Gobie and Cornelius F. Ivory

13. Scale-Up of Isoelectric Focusing 185
    Milan Bier

14. Large-Scale Gel Chromatography: Assessment of Utility for Purification of Protein Products from Microbial Sources..........................193
    James J. Kelley, George Y. Wang, and Henry Y. Wang

15. Electron Paramagnetic Resonance Spectroscopy Studies of Immobilized Monoclonal Antibody Structure and Function........................208
    Erik J. Fernandez, Forrest B. Fernandez, Roger B. Jagoda, and Douglas S. Clark

Author Index..217

Subject Index...217

PREFACE

ONE OF THE MOST DIFFICULT and challenging problems facing large-scale biotechnology today is to find and develop appropriate recovery, separation, and purification processes. The area of large-scale bioseparations is one to which biologists, physical biochemists, and particularly biochemical engineers have important contributions to make. Some of the most recent advances and developments that have already started to find practical applications are

- membrane separations, including the use of membrane bioreactors and liquid emulsion membranes;
- continuous or semicontinuous chromatographic separations, including the use of a number of affinity methods and monoclonal antibodies;
- two-phase extraction processes such as aqueous systems and the use of reverse micelles;
- precipitation techniques;
- electrically driven separation processes;
- methods of product secretion, cell permeation, disruption, and selective enzymatic lysis of microbial cells for intracellular product release;
- product solubilization and renaturation of proteins or polysaccharides present in inclusion bodies or granules.

This book covers several of the emerging areas of separations in biotechnology and is not intended to be a comprehensive handbook. It includes recent advances and latest developments in techniques and operations used for bioproduct recovery in biotechnology and applied to fermentation systems as well as mathematical analysis and modeling of such operations. The topics have been arranged in three sections beginning with product release from the cell and recovery from the bioreactor. This section is followed by one on broader separation and concentration processes, and the final section is on purification operations. The operations covered in these last two sections can be used at a number of different stages in the downstream process.

A crucial question remaining is how to design a flowsheet or product recovery operation sequence. Three main points to keep in mind are (1) integrating recovery with the fermentation system, (2) integrating the different separation and purification stages to design the optimum sequence, and (3) assessing the possibility of a continuous operation.

Revised versions of papers presented in the symposium upon which this book is based as well as papers presented in other sessions that were relevant to the topic have been included in this volume. In addition, we have included a few keynote chapters on areas we felt had not been well covered at the meeting.

We gratefully acknowledge the assistance of many reviewers who helped us with critical and constructive comments on the original manuscripts. We would also like to acknowledge the support and well-organized help of the staff at the ACS Books Department.

JUAN A. ASENJO
Columbia University
New York, NY 10027

JUAN HONG
Illinois Institute of Technology
Chicago, IL 20742

PRODUCT RELEASE AND RECOVERY

1

Protein Release from Chemically Permeabilized *Escherichia coli*

David J. Hettwer and Henry Y. Wang

Department of Chemical Engineering, The University of Michigan, Ann Arbor, MI 48109-2136

> An important factor complicating the recovery of recombinant proteins from Escherichia coli is their intracellular location. An alternative to the commonly used method of releasing these proteins by mechanical disruption is to chemically permeabilize the cells. The objective of this research was to characterize the protein release kinetics and mechanism of a permeabilization process using guanidine-HCl and Triton-X100. The protein release kinetics were determined as a function of the guanidine, Triton, and cell concentrations. Some of the advantages over mechanical disruption include avoidance of extensive fragmentation of the cells and retention of the nucleic acids inside the cell structure.

The recent development of recombinant DNA technology has made it feasible to produce interferon, human growth hormone, insulin, and other proteins in the bacterium Escherichia coli. An important factor complicating the recovery process is the retention of the protein product inside the microbial cell. This has necessitated the development of processes capable of releasing protein from E. coli. Protein release on an industrial scale is commonly achieved by mechanically breaking the cell in a high pressure homogenizer or a ball mill. Disruption in a high pressure homogenizer is caused by pressure gradients established when a pressurized cell suspension is forced through a narrow orifice, whereas with a ball mill, disruption is caused by shear forces generated by grinding the cells with abrasive particles (1).

These mechanically based protein release methods have several undesirable properties. One problem is that extensive fragmentation of the cells makes the subsequent centrifugation difficult (2,3). Adding to the problem of cell fragment removal is the high viscosity imparted to the solution by the released nucleic acids (4). A nucleic acid removal step is necessary to decrease the solution viscosity and avoid potential interference with fractional precipitation and chromatography (5). Another undesirable property is that the harsh action of mechanical disruption causes the release of nearly all the soluble cellular protein. Extensive purification schemes are required to isolate the product from these extraneous cellular proteins.

One alternative to mechanical disruption is to treat the cells with membrane active compounds that can permeabilize the cell to protein without causing extensive breakage of the cell. The objective of this research was to study the protein release kinetics and mechanism of a permeabilization process using guanidine-HCl and Triton-X100. Guanidine-HCl, a chaotropic agent, has been demonstrated to be capable of solubilizing protein from E. coli membrane fragments (6). Presumably, this occurs via guanidine's interaction with water which allows hydrophobic groups to become thermodynamically more stable in an aqueous phase (7). Triton-X100, a nonionic detergent that has a high binding affinity for hydrophobic species, is very effective in binding to and solubilizing phospholipids from E. coli inner membrane and outer wall fragments (8).

Methods

Cell preparation. Escherichia coli K12, strain W3110, was grown in a 14 liter fermenter at 37°C, pH 7.0 using defined media. Additional nitrogen was supplied by NH_4OH which was automatically fed to control the pH. The fermentation broth was harvested in the late exponential phase and cooled to 4°C. The cells were immediately centrifuged at 4°C and washed with buffer (.1M Tris, pH 7.0). Following a second centrifugation, the cells were resuspended in buffer to give a dense cell suspension ($\sim$ 50 g protein/l).

Cell permeabilization. The permeabilization process was started by adding 30 ml of the cell suspension to 70 ml of a buffered solution containing guanidine-HCl and/or Triton X100. The reported concentrations of Triton, guanidine, and cells always correspond to the concentrations after mixing these solutions. The mixture was shaken at 200 rpm in a 4°C incubator. Samples were withdrawn at various times and were immediately centrifuged. The supernatant was assayed to determined the release of the various cell components. Analysis of the pellet was done to perform a mass balance.

Analysis of cell components. Protein was determined with the Bradford dye binding assay using bovine serum albumin as standard (9). Interference by Triton X100 was accounted for by ensuring that every sample had .2% Triton. In order to determine the amount of unreleased protein from the sample pellets, all samples were treated for 5 minutes with 1N NaOH at 100°C.

DNA was determined by the diphenylamine reaction (10). Two 45 minute extractions at 70°C with $.5N$ $HClO_4$ were used to release DNA from the sample pellets. Interference from guanidine was accounted for by making each sample .4M guanidine.

RNA was determined by the orcinol procedure (11). Two 15 minute extractions at 70°C in $.5N$ $HClO_4$ were used to release RNA from the sample pellets. Interference from Triton X100 was accounted for by making each sample 1% Triton.

Results and Discussion

Figure 1 shows the protein, DNA, and RNA release profiles obtained when E. coli cells are mechanically disrupted with .1 mm glass beads. The cell concentration profile, normalized to the initial cell concentration, was obtained with a bacterial counting chamber. The decrease in the cell concentration indicates that extensive fragmentation of the cells is occurring. A nearly mirror image release of DNA, RNA, and protein results as cellular components spill out into the extracellular fluid. The maximum protein release, 70%, is probably indicative of a significant amount of cellular protein being associated with the membrane and wall fragments.

A similar characterization for cells treated with 2M guanidine and 2% Triton is shown in Figure 2. The protein release, based on total cellular protein, levels off at 35%. RNA is released to a lesser extent ($\sim$15%) and very little DNA ($\sim$5%) is released from the cells. The constant cell concentration indicates that the release is not the result of cell fragmentation.

From these results, three major differences between chemical permeabilization and mechanical disruption can be identified. First, the release occurs by fundamentally different mechanisms. With mechanical disruption the cells are essentially torn apart, whereas with chemical treatment the cell structure is still present but has been altered to allow release of intracellular components. Second, there is a nearly complete preferential release of protein over DNA. Third, there is a partial selective release of protein over RNA. This selectivity may result from a molecular sieving mechanism. The average protein molecular weight is 40,000 whereas the cellular DNA has a molecular weight of 2.5 x 10^9 (12). The molecular weight distribution of RNA; 18% is 25,000, 27% is 500,000, and 55% is 1,000,000 is such that most of the RNA is also significantly larger than proteins (12).

These differences suggest several advantages of the chemical permeabilization method. First, avoiding cell breakage should simplify the cell removal step. Second, retention of the nucleic acids inside the cell should eliminate the need for a nucleic acid precipitation step. Another advantage is that the permeabilization process also kills the cells thereby eliminating the need for the federally mandated cell killing step.

Figure 2 showed that $\sim$35% of the total cellular protein is released upon treating the cells with 2M guanidine and 2% Triton. A more complete description of the effect of varying the guanidine and Triton concentrations on the final amount of protein released is shown in Figure 3. Two sets of extractions were conducted: one consisted of using 2% Triton with a range of guanidine concentrations, the other consisted of using 2M guanidine with a range of Triton concentrations. These results indicate that the guanidine-HCl concentration is the more sensitive parameter. Manipulation of the guanidine concentration in the presence of 2% Triton lead to release yields that ranged from 6% to 60% whereas varying the Triton concentration from 0% to 8% in the presence of 2M guanidine only changed the yield from 25% to 40%.

The time profiles of the 2M/2%, 2M, and 2% treatments, shown in Figure 4, indicate a synergistic effect between guanidine and Triton. The protein release profile of the 2M/2% treatment is not simply the

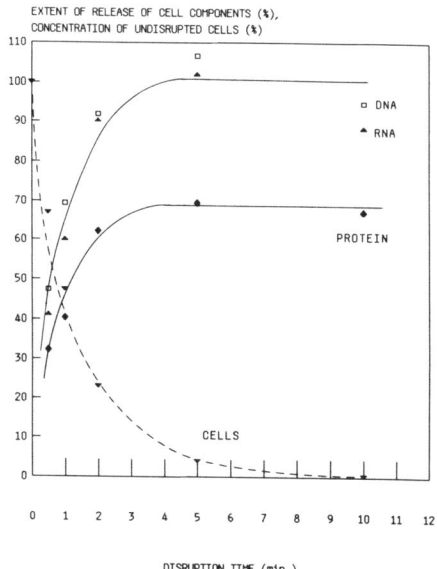

Figure 1. Extent of cell breakage and release of cellular protein, DNA, and RNA during mechanical disruption with .1 mm glass beads.

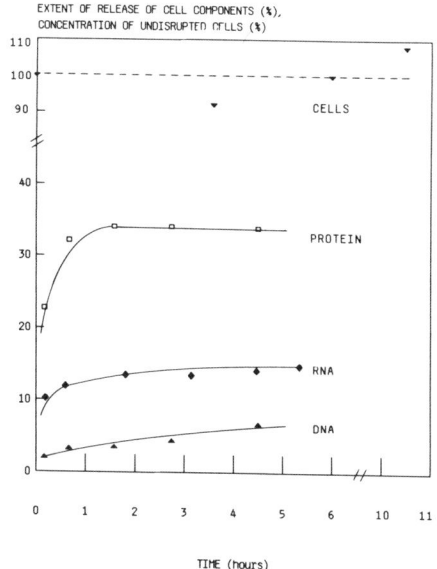

Figure 2. Release of cellular protein, DNA, and RNA during treatment with 2M guanidine HCl and 2% Triton X100.

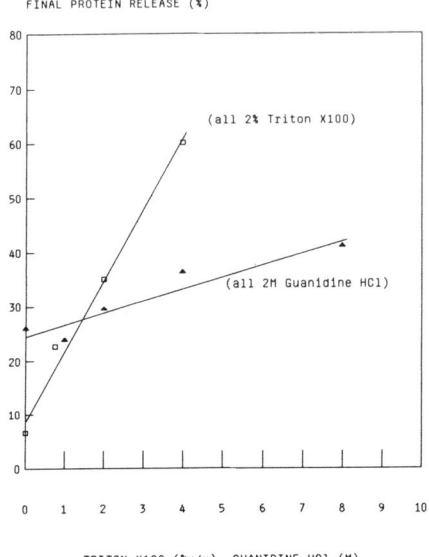

Figure 3. Effect of Triton X100 and guanidine HCl on the protein release yield.

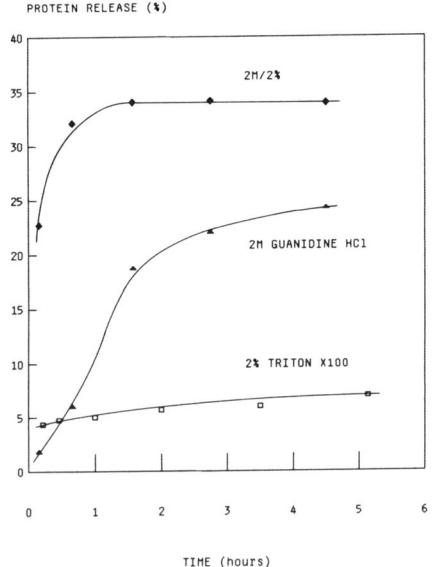

Figure 4. Synergistic effect on the protein release profile between guanidine HCl and Triton X100.

addition of the profiles obtained when 2M guanidine and 2% Triton are used individually. The acceleration of the rate of protein release by Triton may be related to the ability of Triton to solubilize lipid membranes. One would anticipate that the combination of 2M guanidine and 2% Triton alters the E. coli inner membrane and outer wall to a greater extent than either individual treatment, thereby producing a more permeable cell.

The effect of varying the cell concentration on the protein release profile of 2M/2% treatments is shown in Figure 5. The cell concentrations are expressed in terms of the protein concentration of the extraction solution. Although no significant effect was observed on the release profile, the release yield decreased by a factor of two upon increasing the cell concentration from 3.6 g/l to 43.3 g/l. The exact nature of the reason for the decreased yield at high cell concentrations is not known. However, depletion of the guanidine and/or Triton during the process is not occurring, as evidenced by the fact that treating cells a second time with fresh guanidine and Triton does not induce additional release (data not shown). If depletion of the guanidine and/or Triton caused the protein release to cease, one would expect that a second treatment would cause further release of protein from the partially affected or as yet unaffected cells.

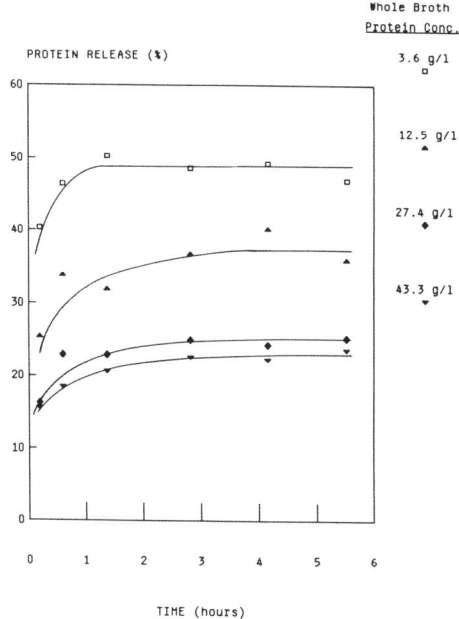

Figure 5. Effect of cell concentration on the protein release profile.

Conclusions

Exposure of E. coli to guanidine-HCl and Triton-X100 induces the release of cellular proteins. The release rate and yield were found to be dependent on the guanidine, Triton, and cell concentrations. Higher concentrations of guanidine and Triton and lower cell concentrations gave greater release rates and yields. Guanidine alone is capable of releasing a significant amount of protein. Triton releases a very low level of protein but substantially increases the rate of release when used in conjunction with guanidine.

The mechanism of the release, a permeabilization of the cell, is fundamentally different from mechanical disruption which involves extensive fragmentation of the cells. The avoidance of extensive cell breakage should simplify the cell removal step and retention of the nucleic acids inside the cell should eliminate the need for a nucleic acid precipitation step. Furthermore, since the treatment kills the cells, a separate cell killing step may be unnecessary.

Acknowledgment

We would like to acknowledge partial support from the National Science Foundation.

Literature Cited

1. Edebo, L. In 'Fermentation Advances'; Perlman, D., Ed.; Academic Press: New York, 1969; p. 249.
2. Schutte, H; Kroner, K. H.; Hustedt, H.; Kula, M. R. Enzyme Microb. Technol. 1983, 5, 143.
3. Bucke, C. In 'Principles of Biotechnology'; Wiseman, A., Ed.; Surrey University Press: New York, 1983; p. 151.
4. Higgins, J. J.; Lewis, D. J.; Daly, W. H.; Mosqueira, F. G.; Dunnill, P.; Lilly, M. D. Biotech. Bioeng. 1978, 20, 159.
5. Wang, D. I. C.; Cooney, C. L.; Demain, A.; Dunnill, P.; Humphrey, A.; Lilly, M. In 'Fermentation and Enzyme Technology'; John Wiley Sons: New York, 1979; Chap. 12.
6. Moldow, C. J. Membrane Biol. 1972, 10, 137.
7. Hatefi, Y.; Hanstein, W. In 'Methods in Enzymology'; Fleischer, S.; Packer, L.; Eds.; Academic Press: New York, 1974; p. 770.
8. Schnaitman, C. J. Bact. 1971, 108(1), 545.
9. Bradford, M. Anal. Biochem. 1976, 72, 248.
10. Burton, K. Biochem. J. 1956, 62, 315.
11. Herbert, D.; Phipps, P. J.; Strange, R. E. In 'Methods in Microbiology'; Norris, J. R.; Ribbons, D. W.; Eds.; Academic Press: London, 1971; p. 210.
12. Brock, T. In 'Biology of Microorganisms'; Prentice-Hall: Englewood Clifts, New Jersey, 1979; p. 131.

RECEIVED March 26, 1986

2
Structured and Simple Models of Enzymatic Lysis and Disruption of Yeast Cells

J. B. Hunter and J. A. Asenjo

Biochemical Engineering Laboratory, Department of Chemical Engineering and Applied Chemistry, Columbia University, New York, NY 10027

Microbial cell-wall-lytic enzymes are widely used in the laboratory for cell breakage, protoplasting of yeasts and bacteria, and for studies of the structure and composition of microbial cell walls (1). Recently lytic systems have come under consideration as a specific and chemically mild way to rupture microbial cells on an industrial scale (2,3). There appear to be attractive commercial applications of lytic systems for the recovery of enzymes, antigens and other recombinant products accumulated within cells, for upgrading of microbial biomass for food and feed uses (4,5) and for the manufacture of functional biopolymers from cell wall carbohydrates (6).
This paper presents two models of enzymatic lysis of yeast cells; a simplified two-step model, accounting for protein release at cell lysis followed by proteolysis, and a more complex mechanistic model which describes the removal of the two layers of the yeast wall and the extrusion and rupture of the protoplast and organelles. The use of these models in predicting the release and breakdown of microbial proteins, and the application of the structured model to enzyme recovery will also be discussed.

One problem in production of recombinant proteins is recovery of the finished product from the cells which accumulate it. This problem is particularly acute in the case of yeasts and fungi, which have tough, thick cell walls which are difficult to rupture mechanically or by sonication. Product secretion is not always feasible, even for low-molecular-weight products, although a newly developed secretion process for yeast (7) appears promising.

There are numerous examples of overproduced recombinant proteins which precipitate intracellularly in E. coli, forming dense inclusion bodies (8); these products include insulin and somatostatin, both very small proteins. In yeast, recombinant viral surface antigen proteins are not secreted, but assemble into particles (9). Subcellular structures such as mitochondria, lysosomes or the vacuole must also be recovered by cell breakage, for use either as biocatalysts (10) or as an initial step in the purification of enzymes associated with such structures. Until now, these products have generally been harvested by mechanically rupturing the cells in a homogenizer, bead mill or French press. The high shear fields, elevated temperatures and gas-liquid interfaces generated in these devices can denature proteins, especially multi-enzyme complexes and membrane-linked proteins (11). Moreover, the separation of cell debris from the products is especially complicated if the product is particulate, fragile or membrane-associated.

Lytic enzyme systems provide a chemically mild, low-shear and catalytically specific alternative to mechanical cell disruption. Depending on the particular lytic system employed and its purity, the enzymes may be engineered to attack cell wall components alone, without product damage. The enzyme lysozyme, active against some bacterial cell walls, has been used to harvest bovine growth hormone granules from E. coli (8), and a membrane-associated hydroxylase complex from P. putida (11); use of other bacteriolytic enzymes from a variety of microbial sources have also been reported (3).

Investigations into the subcellular location of enzyme activities in microbial cells suggest that one or more enzyme products could be specifically fractionated from a single batch of cells by properly controlling cell disruption. Invertase in yeast is possibly the best example of this principle. The studies leading to discovery of its location (in the periplasmic space) have been summarized by Phaff (12; p.171-173), and a sample process for its recovery has been proposed (4). The recovery of several different enzymes in high yield and high relative purity should be possible using a combination of lytic enzymes, surfactants and osmotic support buffers to selectively and sequentially release proteins from particular structures.

Cell fractionation by mechanical rupture has already come under investigation. Two separate studies of mechanical rupture of yeast showed different rates of release for enzymes in different cell locations (13,14). Wall-linked and periplasmic enzymes were released relatively faster than total protein, soluble cytoplasmic enzymes at about the same rate, and the mitochondrial enzyme fumarase later than total protein (13). Proteolysis by the yeast's own enzymes was not found to be a problem. Activities of the released enzymes declined slowly or not at all when disruption was continued after the end of protein release, and the effect of shear was not separated from the effect of proteolysis. Shetty and Kinsella (15) also found a low rate of proteolysis after mechanical disruption, though thiol reagents added to weaken the cell walls before disruption caused an important increase in the extent of protein breakdown.

Model Background

Yeast cell structure. The extensive body of literature on cell wall composition and structure has recently been reviewed by Ballou (16) and earlier by Phaff (12).

As an engineering approximation, the cell wall of yeast may be considered as a two-layer structure. (Figure 1) The inner wall is composed of a mixture of branched β(1-3) and β(1-6) linked glucans, glucose polymers similar to cellulose (12). The outside of the glucan layer is covered with a mannan-protein complex consisting of a cross-linked network of protein molecules, to which are attached two types of mannan: an acidic oligosaccharide, and a higher molecular weight phosphomannan having a d.p. of about 100 (17). From the perspective of cell lysis, this mannoprotein layer serves to protect the glucans from hydrolytic enzymes (18,19,20). Within the two wall layers is the protoplast, comprised of a plasma membrane enclosing the cytosol and the subcellular structures.

Enzymes of the lytic system. Microbial yeast-lytic enzyme systems are widely distributed in nature, and have been isolated from Rhizoctonia sp., (4), Bacilus circulans (21), Coprinus macrorhizus (22), and Cytophaga sp. (23), among other sources.

Crude yeast lytic enzyme systems comprise several hydrolytic activities, often including chitinase, mannanase, and a variety of proteases and glucanases (1). Only two of these activities, a lytic protease and a lytic glucanase, are essential for lysis (19,24,20). Lytic glucanases usually bind preferentially to long chains of β(1-3) glycosidic linkages, such as those found in microfibrillar yeast wall glucan. In general, the lytic glucanases have an endo- action pattern but some attack exo-wise, releasing oligosaccharides of 5 glucose units from the structural yeast glucan. Other glucanases, with different substrate specificity and action patterns, are usually present in the lytic system and act synergistically to degrade insoluble yeast glucan to glucose and disaccharides (25). Lytic proteases have a characteristic high affinity for the yeast wall surface, and often have anomalously low activities against conventional protein substrates. Their role in lysis of viable yeast cells cannot be substituted by ordinary proteases. (20,26).

We used a lytic system from Oerskovia xanthineolytica LL-G109 from the collection of M. Lechevalier, at Rutgers University. Filtered culture broth was used as the enzyme source. Details of the enzyme production are given elsewhere (27,28). The lytic activity of the Oerskovia system is due to a lytic protease and an endo β(1,3) glucanase (20), possibly supplemented with an exo β(1-3) glucanase removing a 5-sugar unit from the chain (29).

Sequence of cell lysis. Enzymatic cell lysis begins with binding of the lytic protease to the outer mannoprotein layer of the wall. The protease opens up the protein structure, releasing wall proteins and mannans, and exposing the glucan surface below (Figure 2). Next, the glucanase attacks the inner wall and solubilizes the glucan (19). When the combined action of protease and glucanase has opened a sufficiently large hole in the cell wall, the plasma membrane and its contents are extruded as a protoplast (1). In osmotically supported buffers containing .55 to 1.2M sucrose or

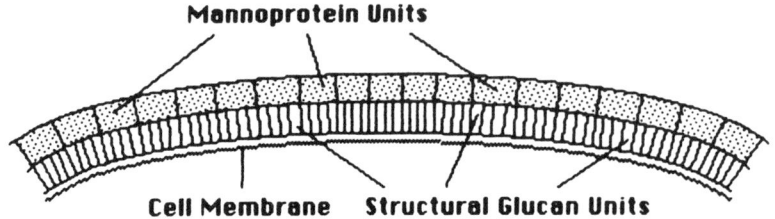

Figure 1 Double-layered structure of the yeast wall, enclosing the cell membrane

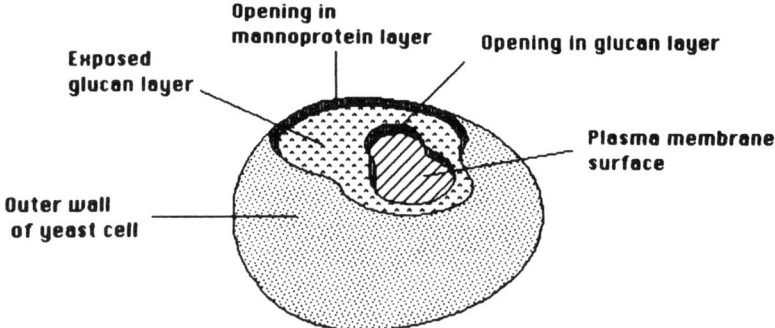

Figure 2 Schematic of lysing yeast cell

mannitol, the protoplast remains intact but in dilute buffers it lyses immediately, releasing cytoplasmic proteins and the organelles which may themselves lyse.

Meanwhile, proteins released from the wall and the cytoplasm are subject to attack by product-degrading protease contaminants in the lytic system (28,30).

Models

Mathematical models with different levels of structure are useful for the design of reactors, to carry out simulation studies, for process optimization and for increasing our understanding of the mechanistic, biological behavior of biochemical systems.

Historically there has been little published work on models of microbial cell lysis. The models proposed for overall cell lysis have been elementary and their application has been limited. First-order and Michaelis-Menten models have been used to estimate the performance of a sample lysis process (2,3). Lysis of freeze-dried Micrococcus lysodeikticus cells by lysozyme was modeled with a second-order rate expression (31). At the other end of the spectrum of mathematical complexity is a model of lysozyme-catalyzed degradation of soluble bacterial cell-wall oligosaccharides, focusing on the degree of polymerization of the substrate and the binding modes of enzyme to substrates (32). Accounting for one enzyme and carbohydrate oligomers up to d.p. 9, it has nine differential equations and ten parameters, and was tested on purified radiolabeled oligosaccharides. Although useful for elucidating enzyme action patterns, such models are too detailed to be readily applied to a multi-enzyme, multi-substrate system.

The two models of yeast lysis presented here have been developed to serve two different purposes. The simple model is a lumped, two-step model which follows the major features of the data and may prove useful for design of lysis reactors. The structured model, which can account for the source of protein within the cell, was developed to gain a mechanistic basis for predicting the effects of untested process conditions, and to aid insight into the physical processes at work during lysis.

Simple model. The simple model was built for compact description of the data in a pre-determined range of yeast and enzyme concentrations. It treats cell lysis and proteolysis as single-step reactions in sequence. Both reactions are modeled with Michaelis-Menten kinetics, even though yeast, the substrate of the first reaction, is particulate and the proteins are soluble. The different enzymes of the lytic system are grouped together into an all-inclusive single enzyme, E, bearing both the proteolytic and yeast-lytic activities. All of the cell structures are also considered together as a unified yeast cell mass, Y.

When a cell is attacked by enzymes it is presumed to dissolve instantaneously, releasing its entire mass as soluble proteins, peptides and carbohydrates. The assumption of instantaneous solution of the cell mass constrains the model for use where the lysis medium is hypo-osmotic and protoplasts cannot survive intact.

Only two independent variables are used: yeast (Y) and enzyme (E); the measured variables are yeast, TCA-insoluble protein (P), TCA-soluble protein (peptides, S), and carbohydrates (C); all are expressed as g/l dry basis. Enzyme concentration was expressed as the volume per cent of crude lytic enzyme preparation added to the reaction mixture. Proteolytic and other causes for lytic enzyme deactivation (e.g., thermal) have been assumed to be negligible (28).

$$\frac{dY}{dt} = -k_a(Y-Y_\infty) - \frac{k_r E \cdot (Y - Y_\infty)}{(Y - Y_\infty) + K_m} \tag{1}$$

$$\frac{dP}{dt} = -f_{py}\left(\frac{dY}{dt}\right) - \frac{k_p E \cdot P}{P + S + K_{mp}\left(1 + \frac{Y}{K_i}\right)} \tag{2}$$

$$\frac{dS}{dt} = -f_{sy}\left(\frac{dY}{dt}\right) + \frac{k_p E \cdot P}{P + S + K_{mp}\left(1 + \frac{Y}{K_i}\right)} \tag{3}$$

$$\frac{dC}{dt} = -f_{cy}\left(\frac{dY}{dt}\right) \tag{4}$$

Variable names and parameter values are given in Table I.

On the right-hand side of equation 1, the initial term represents autolysis and the second term, enzymatic lysis. Equation 2 describes protein breakdown by product-degrading proteases. The first term on the right side stands for the protein released from lysing cells, and the second term, breakdown of the protein already in solution. Equation 3 shows that peptides are released from lysing yeast, but also arise from breakdown of longer proteins, P. Since the protease activity against soluble proteins is considered non-specific, both long- and short-chain proteins will be attacked by the enzyme with essentially the same affinity per gram of substrate. Hence, S will act as a competitive inhibitor of the enzyme activity against P, where the inhibition constant is equal to the Michaelis constant K_{mp}. Carbohydrate release is shown in equation 4.

Parameters for the simple model were determined graphically by Eadie-Hofstee plotting of initial reaction rates and substrate concentrations. Details are given elsewhere (30). As has been observed in hydrolysis of other solid substrates, a residue of non-lysed substrate was found at extended reaction times, when dY/dt tended toward zero. The extent of reaction was strongly dependent on initial substrate and enzyme concentrations (33,34). An empirical funciton for Y_∞ was fitted to the ultimate turbidity data for lysis runs at a variety of initial yeast and enzyme concentrations using a least squares method. The calculated values for Y_∞ were used in the simulations (30). Figure 3 shows results of the simple model.

Structured model. This model considers lysis of the cell from the viewpoint of progressive breakdown of the cell structures, starting from the outer wall layer and progressing to the subcellular structures inside the protoplast (35). Here the cell is divided into

Table I. Lumped model variables and parameters

Variables - Simple Model

Y	Yeast, mg/l
Y_O	Original yeast concentration
Y_∞	Ultimate yeast concentration, mg/l; proportional to residual turbidity.

$$\frac{Y_\infty}{Y_O} = a + bE + cY_O + \frac{d}{Y_O}$$

P	Protein (TCA-insoluble), mg/l
S	Peptides (TCA-soluble), mg/l
C	Carbohydrates, mg/l
E	Enzyme, % (v/v) of reaction mixture

Parameters-Simple Model

Y_∞ constants:
- a: 3.6342×10^{-1}
- b: -2.6584×10^{-3}
- c: 6.0442×10^{-6}
- d: -9.9603×10^{1}

k_a	Rate constant for autolysis	$3.987 \times 10^{-4} \text{min}^{-1}$
k_r	Rate constant for lysis, simple model	15.51 mg/L-min-%ez
K_m	Michaelis constant for lysis,	1902 mg/L
k_p	Rate constant for proteolysis,	4.441 mg/L-min-%ez
K_{mp}	Michaelis constant, proteolysis,	4598 mg/L
K_i	Inhibition constant, proteolysis,	26077 mg yeast/L
f_{py}	Fraction of protein in yeast	0.4048
f_{sy}	Fraction of peptides in yeast	0.0777
f_{cy}	Fraction of carbohydrates in yeast	0.3687

four regions; the outer wall or wall protein (WP); inner wall or wall glucan (WG); the cytosol (CS) and the organelles, here grouped together as mitochondria (MI). The lytic system is approximated as three enzymes, a lytic glucanase, E_g, which hydrolyzes the inner cell wall glucan, a lytic protease, E_p, which attacks only the outer wall layer and a destructive protease, E_d, active against soluble proteins. Product inhibition is included in all enzyme reactions. Adsorption and desorption of the enzymes to the yeast wall is neglected, since adsorption kinetics appeared instantaneous on the time scale of our measurements (35). A schematic of the reaction pathways is shown in Figure 4.

Special variables.

$$EGA = (WG - r \cdot WP)$$

The glucan hydrolysis rate is not related directly to total glucan concentration WG, but rather to the amount of glucan made accessible to attack through removal of wall protein from the outside of the cell. EGA, "exposed glucan, accessible" represents the amount of glucan uncovered by removal of the outer wall. The proportionality constant r is the weight ratio of wall glucan to wall protein.

$$PBR = k_p \cdot E_d \frac{\left(\frac{P}{K_{mp}}\right)}{\left(1 + \frac{P + S}{K_{mp}}\right)}$$

The overall rate of soluble protein hydrolysis, PBR, protein breakdown rate, accounts for destruction of soluble protein by the destructive protease.

The release of cytosol into the medium depends on the osmotic breakage of the protoplasts, which occurs at a rate approximately proportional to the osmotic gradient across the plasma membrane (36). The internal osmolality of the cells was estimated to be 0.617 Os/L (35), where 1 Os/L is equivalent to 1 Mol/L of an ideal solute. The external osmolality is the sum of the contribution from the buffer system in the medium (about 0.02M in our experiments) and the substances released by lysing protoplasts. The stabilization of the remaining cells by these substances is far stronger than could be expected solely on the basis of osmotic effects, and could result from the release of cations which interact with specific receptors on the plasma membrane (37). The release of soluble products of glucan and protein hydrolysis are also expected to add to the stabilizing effect of the lysate.

The effective osmolality of cell lysate was fit to a Langmuir expression, where OSM_L is the maximum stabilizing effect and K_{osm} is the equilibrium constant for interaction of the stabilizers with the protoplasts. The resulting equation,

$$OSM_x = B_o + \frac{OSM_L \cdot K_{osm}(CS^* + CS_o - CS)}{1 + K_{osm}(CS^* + CS_o - CS)}$$

expresses total effective osmolality in the lysis medium. B_o is the original osmolality of the lysis buffer and CS^* is the sum of protein, peptides and carbohydrates present at the start of reaction.

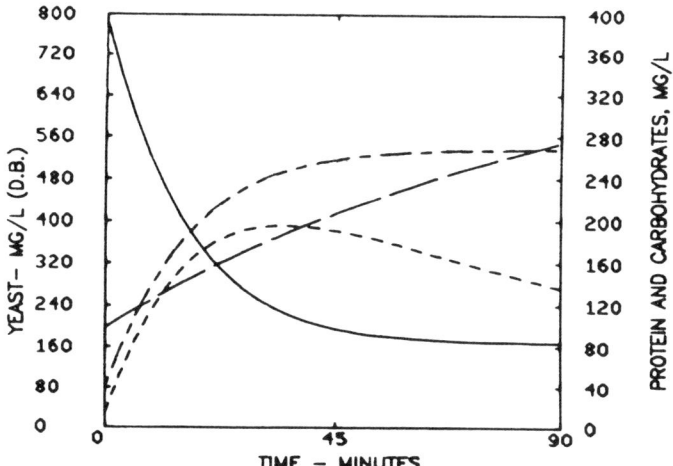

Figure 3 Simple model simulation of yeast lysis
——————— Yeast cell mass, mg/l – – – – – Protein, mg/l
— — — Peptides, mg/l   — – — Carbohydrates, mg/l
0.78 g/l yeast concentration; 10% enzyme

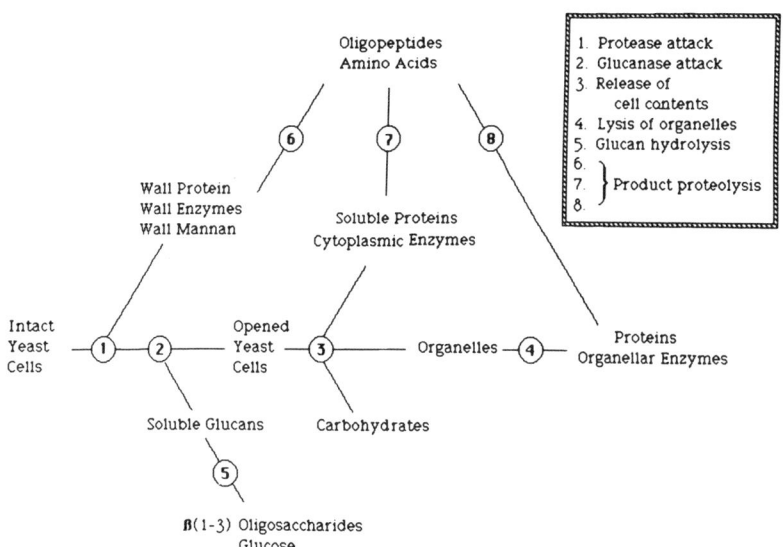

Figure 4 Reaction pathways for structured model

Based on the product $OSM_L \cdot K_{osm}$, the stabilizing effect of cell lysate at low concentrations is equivalent to 4.4×10^{-4} Os/mg cytosol released (35).

Wall hydrolysis equations.

$$\frac{d(WP)}{dt} = \frac{-k_{wp} E_p \left(\frac{WP}{Km_{wp}}\right)}{1 + \frac{WP}{Km_{wp}} + \frac{WP_o - WP}{Ki_{wp}}} \qquad (1)$$

$$\frac{d(WG)}{dt} = \frac{-k_{wg} E_g \left(\frac{EGA}{Km_{wg}}\right)}{1 + \frac{EGA}{Km_{wg}} + \frac{WG_o - WG}{Km_{sg}}} \qquad (2)$$

Release of cytosol and mitochondria. The osmotic gradient between protoplasts and buffer or mitochondria and buffer drives the release of protein into the medium. If the osmolality of the external medium exceeds the internal osmolality of the protoplast or organelle, no rupture occurs. The osmolality decreases internally, and increases externally, as material is released from the protoplast. In addition, the release of cytosol is proportional to the size of the opening in the wall glucan, up to a maximum hole size of 1/3 of the cell's surface area.

$$\frac{d(CS)}{dt} = -(CS) \cdot k_a -$$
$$(CS) \cdot k_r [\max(0, OSM_i \cdot \left(\frac{CS}{CS_o}\right) - OSM_x)] \cdot \max(.33, 1 - \frac{WG}{WG_o}) \qquad (3)$$

$$\frac{d(MI)}{dt} = -\left(\frac{MI_o}{CS_o}\right) \cdot \frac{d(CS)}{dt} - k_{rm}[\max(0, 0.3 - OSM_x)] \cdot MI \qquad (4)$$

Soluble products. Values for TCA-insoluble protein, peptides and carbohydrates released were estimated by summing the contribution to each pool from the breakdown of each cellular structure.

$$\frac{dP}{dt} = -f_{pwp} \cdot \left(\frac{d(WP)}{dt}\right) - f_{pcs} \cdot \left(\frac{d(CS)}{dt}\right) +$$
$$f_{pm} \cdot k_{rm}[\max(0, 0.3 - OSM_x)] \cdot MI - PBR \qquad (5)$$

$$\frac{dS}{dt} = -f_{swp} \cdot \left(\frac{d(WP)}{dt}\right) - f_{scs} \cdot \left(\frac{d(CS)}{dt}\right) +$$
$$f_{sm} \cdot k_{rm}[\max(0, 0.3 - OSM_x)] \cdot MI + PBR \quad (6)$$

$$\frac{dC}{dt} = -\frac{d(WG)}{dt} - f_{ccs} \cdot \frac{d(CS)}{dt} \quad (7)$$

Total yeast cell mass was estimate as the sum of WG, WP, CS, and MI (structures remaining with the cell), with an added factor accounting for non-protein, non-carbohydrate substances in the cell. These sums generate values for yeast, protein, peptides and carbohydrates for comparison to experimental measurements.

The variables for the structured model are listed in Table II. Parameter values are given in Table III (35).

Table II. Structured Model Variables

EGA	Exposed glucan, accessible for hydrolysis by glucanase
Y_o	Initial quantity of yeast, mg/l dry basis
WG	Wall glucan, mg/l
WG_o	Original amount of glucan; = $Y_o \cdot fWG_o$, mg/l
CS	Cytosol, mg/l
CS_o	Original quantity of cytosol = $Y_o \cdot fCS_o$, mg/l
CS*	Initial amount of osmotically stabilizing material in reaction medium: $P_o + S_o + C_o$
WP	Wall protein, mg/l
P	Long-chain protein (TCA - insoluble), mg/l
S	Oligopeptides (TCA - soluble), mg/l
M	Mitochondrial mass, mg/l
M_o	Original mass of mitochondria in cell = $Y_o \cdot fM_o$, mg/l
B_o	Osmotic strength of buffer, Os/kg
E_g	Glucanase enzyme of lytic system, %(V/V) of mixture
E_p	Lytic protease enzyme, %(V/V) of mixture
E_d	Destructive or product-degrading protease, %(V/V)
WE	Yeast enzyme in walls, mg/l
CE	Yeast enzyme in cytosol, mg/l
ME	Yeast enzyme in mitochondria, mg/l

Table III. Structured model parameters and their values

r	Ratio of wall glucan to wall protein	1.326 mg WG/mg WP
k_{wg}	Rate constant, glucanase on WG	10.58 mg/L-min-%ez
Km_{wg}	Michaelis constant, glucanase on WG	4424 mg/L
Km_{sg}	Michaelis constant, glucanase on soluble glucan	800 mg/L
k_{wp}	Rate constant, proteolysis of WP	4.441 mg/L-min-%ez
Km_{wp}	Michaelis constant	459.8 mg/1
Ki_{wp}	Inhibition constant, proteolysis of WP	919.6 mg/1
k_p	Rate constant, proteolysis of P	1.441 mg/L-min-%ez
K_{mp}	Michaelis constant	4598 mg/L
k_a	Rate constant for CS leakage	3.987×10^{-4} min^{-1}
k_r	Rate constant for CS release	1.6667 min^{-1}
k_{rm}	Rate constant for MI breakage	0.6 min^{-1}
f_{pwp}	Fraction protein in wall protein	0.9434
f_{pcs}	Fraction protein in cytosol	0.3753
f_{pm}	Fraction protein in mitochondria	0.75
f_{swp}	Fraction peptides in wall protein	0.0566
f_{scs}	Fraction peptides in cytosol	0.1170
f_{sm}	Fraction peptides in mitochondria	0.05
f_{ccs}	Fraction of carbohydrates in cytosol	0.3145
fCS_o	Initial fraction of cytosol in yeast	0.5612
fWG_o	Initial fraction of wall glucan	0.1922
fWP_o	Initial fraction of wall protein	0.1450
fM_o	Initial fraction of mitochondria	0.7652
OSM_i	Internal osmolality of yeast cell	0.617 Os/L
OSM_L	Maximum effective osmolality of released lysis products	0.539 Os
K_{osm}	Equilibrium constant for osmotic stabilization of protoplasts	8.135×10^{-4} L/mg CS
f_{ewp}	Proportion of enzyme in wall protein	0.01
f_{ecs}	Proportion of enzyme in cytosol	0.01
f_{emi}	Proportion of enzyme in mitochondria	0.01

The structured model simulates the progress of lysis in terms of the cell's structural components during lysis. The decrease in WP starts immediately, as it is the first component attacked by the lytic enzymes. WG breakdown lags WP removal, and cytosol release lags glucan breakdown, as suggested by the sequentiality built into the model. The mitochondria are released last, and tend to accumulate because they are more resistant to osmotic rupture than the protoplasts. A typical graph is shown in Figure 5a. In Figure 5b, structured model estimates of yeast cell mass, protein, peptides and carbohydrates are presented for the same enzyme and yeast concentrations.

Results

The simple and structured model simulations for yeast mass and soluble protein, peptides and carbohydrates are compared in Figure 6 for the yeast and enzyme concentration shown in Figures 3 and 4, and in Figure 7 for a concentrated yeast cell slurry. The simple model fits the data fairly well at both yeast concentrations, in every variable except the peptides. The fit for all variables at longer reaction times is directly related to use of the extent-of-reaction term Y_∞ in the yeast lysis equation.

The structured model provides a distinct improvement over the simple model, in the initial stages of the reaction. The initial lags in the hydrolysis of total yeast mass and carbohydrate in figures 6 and 7 are very well represented. The possibility remains that the initial lags relate partly to adsorption of lytic enzymes to the cell wall. On the time scale of our experiments, however, adsorption appeared to be instantaneous (35).

At high yeast concentration (figure 7) at the later stages of reaction, the carbohydrates continue to rise though turbidity is levelling off. Apparently some wall hydrolysis is occurring even though the total yeast solids concentration is not visibly declining. This result is in contrast to figure 6, where the structured model follows carbohydrate data closley, and the hydrolysis of wall glucan is estimated to go essentially to completion. Presumably the glucanase attacks the more susceptible amorphous glucan at a higher rate than the fibrillar glucan fraction of the wall. Such dependence on physical structure is well known to occur in enzymatic hydrolysis of cellulose (38,34). If the analogy is correct, the amorphous carbohydrateds could be solubilized without substantially changing the microfibril network structure in the wall, or releasing protoplasts. Cellulose/cellulase system results also suggest that this effect ought to be more pronounced at the higher yeast-to-enzyme ratio shown in Figure 7, than at the lower ratio of Figure 6.

A limitation of both models is overestimation of the amount of protein released. Peptide predictions by the simple model are too high as well. The effect resembles a gap in the material balance, as if the models predict a larger quantity of proteinacious material than is actually present in the cells. Possibly some cytoplasmic proteins are not released during protoplast breakage. In figure 6, insoluble proteins could account for a substantial part of the residual yeast at 90 minutes digestion.

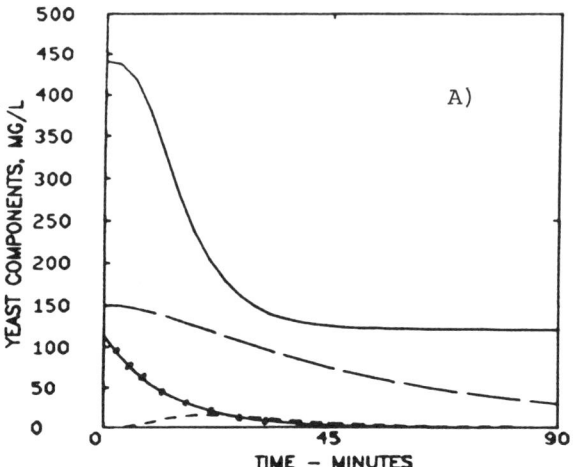

Figure 5   Structured model simulation of yeast lysis

   5a Cell structures

——————— CY cytosol   — — — WG wall glucan

•—•—•—• WP wall protein   - - - - MI mitochondria

mg/l

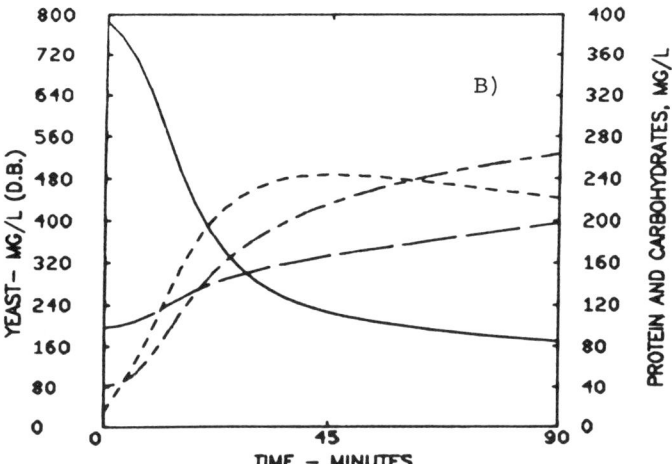

   5b Cell mass and released compounds

——————— Yeast cell mass, mg/l   - - - - Protein, mg/l

—— —— Peptides, mg/l   —— - —— Carbohydrates, mg/l

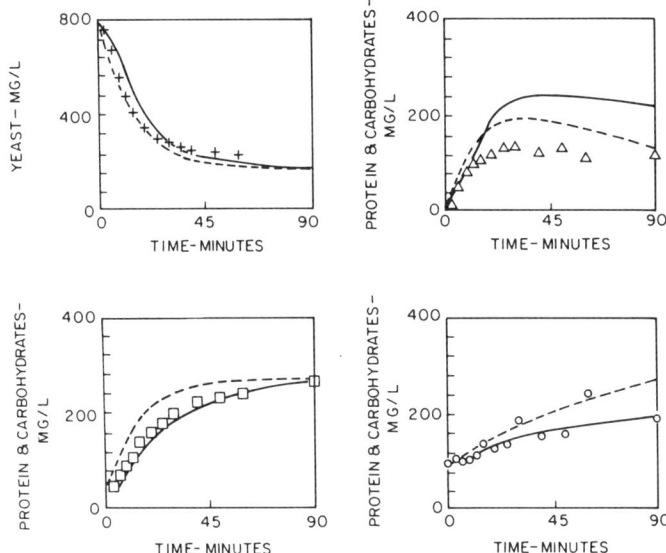

Figure 6 Comparison of simple and structured models - Intitial yeast concentration 0.78 g/l (d.b), 10% enzyme
———— Structured model   — — — — Simple model

+ Yeast △ Protein
□ Carbohydrates ○ Peptides

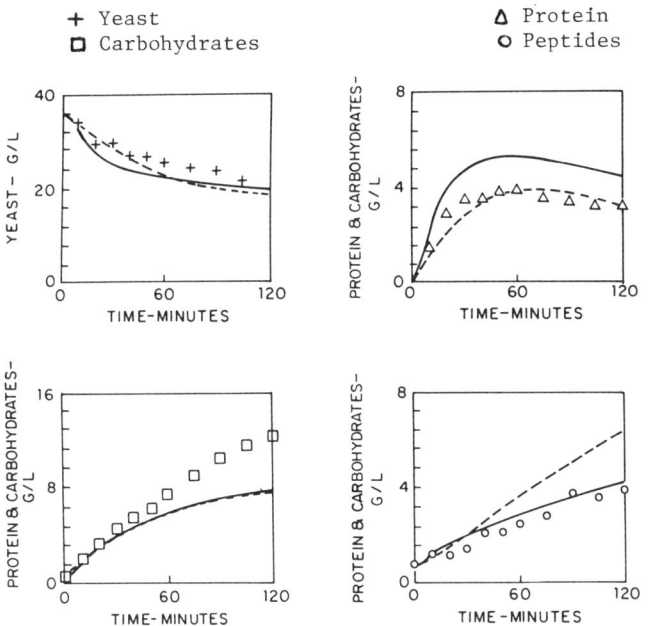

Figure 7 Comparison of simple and structured models - Initial yeast concentration 36.3 g/l (d.b) 40% enzyme Symbols as in figure 6

Work continues in two areas: purification of the lytic system to allow protease and glucanase levels to be controlled independently and investigation of the release of site-specific yeast enzymes and subcellular fractions by enzymatic lysis.

Applications

The structured model's detailed accounting of the fate of cell structures can be used to make predictions about the effects of a number of important process variables, for example:

- The ratio of lytic protease to glucanase in the lytic system
- The effect of pH or temperature on synergism between the lytic enzymes
- Elimination of "destructive" protease (for bioactive protein recovery) or supplementation with additional proteases (for food and feed applicaitons)
- The addition of protease inhibitors at a point or points during lysis, effectively lowering k_p or increasing K_{mp}.
- Osmotic buffering strategies for recovery of bioactive protein from different sites in the cell.

Cell Fractionation Simulation. The wall protein, cytosol and organelles of yeast each contain enzymes which are found nowhere else in the cell. Some examples of these enzymes include invertase in the walls, glycolytic pathway enzymes in the cytosol and fumarase in the mitochondria (13). A model of recovery of these enzymes is offered here.

Enzyme accumulation. For the purpose of simulation, wall-linked and periplasmic enzymes (WE) are considered to be a part of the outer wall protein. Cytoplasmic and mitochondrial enzymes (CE,ME) are assumed to be some fraction of the cytoplasmic and mitochondrial mass, respectively. The equations describing their release and hydrolysis are exactly analogous to equation 6 for total long-chain protein.

$$\frac{d(WE)}{dt} = -f_{ewp} \cdot \left(\frac{d(WP)}{dt}\right) - \left(\frac{(WE)}{P}\right) \cdot PBR \qquad (8)$$

$$\frac{d(CE)}{dt} = -f_{ecs} \cdot \left(\frac{d(CS)}{dt}\right) - \left(\frac{CE}{P}\right) \cdot PBR \qquad (9)$$

$$\frac{d(ME)}{dt} = f_{em} \cdot k_{rm} \; [\max(0, 0.3-OSM_x)] \cdot MI - \left(\frac{ME}{P}\right) \cdot PBR \qquad (10)$$

Variables and parameters are included in Table III.

Figure 8 shows a simulation of enzyme recovery from the wall, cytosol and mitochondria. The concentrations of recoverable enzyme are normalized to the initial amount of enzyme present in the cell site. The curves rise as enzyme is released from a site, then fall as it is hydrolyzed. It may be seen that the lytic system is usable even as a crude preparation to recover wall linked yeast enzymes in 60 to 80% yield. The yield of yeast wall enzyme depends on

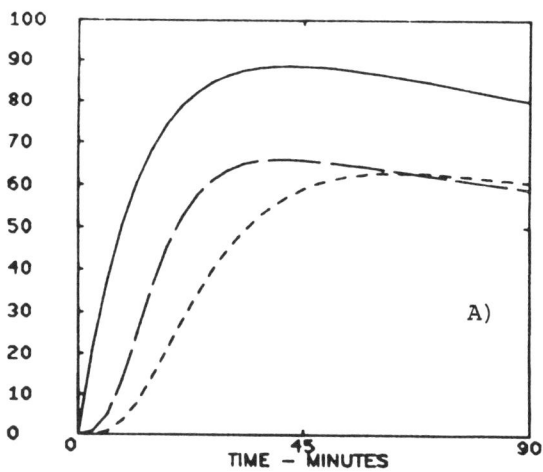

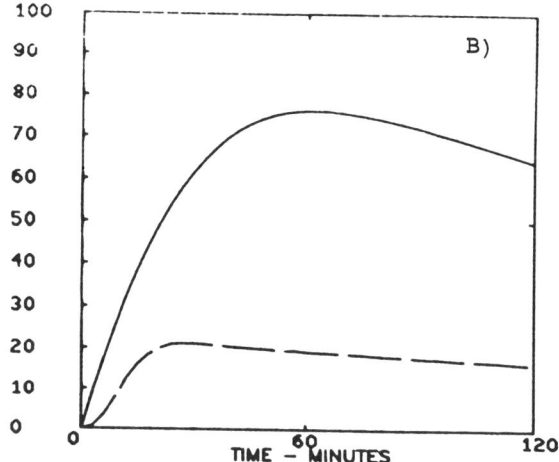

Figure 8   Release of site - specific enzymes - Simulation

    8a   Initial yeast concentration, 0.78 g/l
    8b   Initial yeast concentration, 36.33 g/l
    ————— Wall enzyme   — — Cytoplasmic enzyme
    - - - - - Mitochondrial enzyme

protease activity and is therefore directly related to the quantity of "destructive" protease present in the lytic system. The product purity depends on the release of proteins from other cell sites, and hence on osmotic factors. At the higher yeast concentration (Fig. 8b), few of the protoplasts and none of the mitochondria release their enzymes into solution. While the yield of wall enzyme is not as good as in Figure 8a, the purity is far higher. With proper osmotic support during lysis, and breakage of osmotically stable protoplasts by mechanical means, the wall, cytoplasmic and mitochondrial fractions can be obtained separately.

A simulation of site-linked product recovery is presented in Figure 9 and Tables IV and V. The calculations assume that site-linked enzymes WE, CE, and ME constitute 1% of the wall protein, cytosol and mitochondria respectively. In the first lysis step, using 20% lytic enzyme broth and osmotic support, 93% of the wall protein (and wall enzyme) is released from the cell wall. Some is hydrolyzed by the "destructive" protease, but 73.8% of it survives to be recovered at the end of the first hour. Since only 3% of the protoplasts burst during this step, little cytoplasm is released. The protein concentration in solution at the end of the hour is 3.83 g/l of which about 1% is WE. If the cells were broken mechanically, the wall enzyme would constitute only 0.35% of the total protein, even assuming that it could be completely solubilized. The simulation also shows a ratio of WE to CE of 7.8 in the medium at the end of the first step, which compares to a ratio of 0.258 on a total-cell basis.

Table IV. Process conditions for enzyme release simulation

First lysis step:

Yeast	36.33g
Enzyme	40%
Buffer	0.3 Os/L
Total volume	1 L

Reaction mixture centrifuged; 5% of supernatant and 100% of pellet retained and resuspended in twice the initial volume of enzyme/buffer solution.

Second lysis step:

Digested yeast:	26.90g
Enzyme	20%
Buffer	0.3 Os/L
Total volume	2 L

Breakage: by stirring or passage through a pump

Protoplast rupture	95%
Mitochondrial rupture	0%

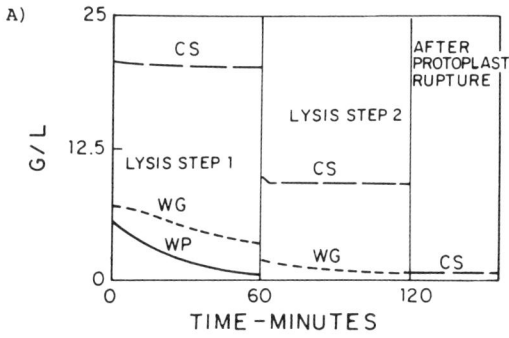

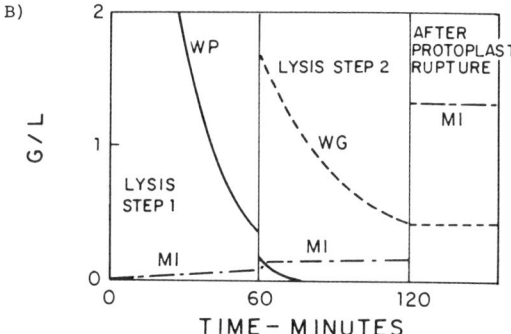

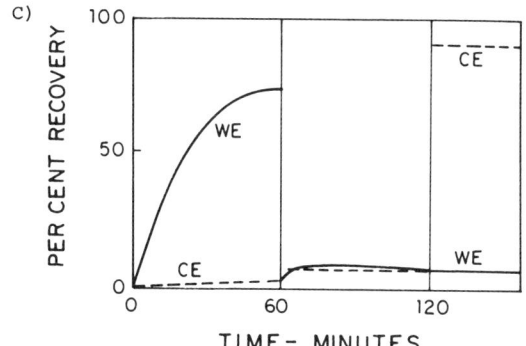

Figure 9   Enzyme recovery from subcellular structures

9a   Cell structure breakdown
——————Wall protein   - - - - - Wall glucan
— — Cytosol   — · —Mitochondria

9b   Cell structure: Close-up, showing mitochondria
9c   Enzyme release, per cent of original enzyme in cell
—————— Wall enzyme   - - - - -Cytoplasmic enzyme

Table V. Yields

		Initial	End of first step	Start of second step	End of second step	After protoplast rupture
Yeast solids	mg/l	36332	26902	26902	21660	4766
Soluble protein	"	8.1	3831.5	191.6	966.9	7307.3
Soluble peptides	"	760.7	2135.0	106.7	684.9	2661.5
Soluble carbohydrates	"	655.0	4449.8	222.5	3321.4	8634.7
Wall protein	"	5268.2	371.0	371.0	0.028	0.028
Wall glucan	"	6983.0	3380.2	908.9	908.9	908.9
Cytosol	"	20389	19779	19779	17783	889.2
Mitochondria (internal)	"	2780.1	2696.9	2696.9	2424.7	121.0
Mitochondria (released)	"	0	83.2	83.2	355.4	2659.1
Wall enzyme	mg/l	0	38.6	.97	4.18	4.18
Cytoplasmic enzyme	"	0	4.95	.12	15.24	184.2
Mitochondrial enzyme	"	0	0	0	0	0

The second digestion was included to decrease the amount of structural glucan from about 50% to about 13% of its original mass, in order to make the cells more fragile, thus easier to rupture mechanically. Only a small amount of cytoplasmic protein is released from the protoplasts during this time - a desirable result, since protein sequestered inside the protoplasts is not attacked by protease.

At the end of the second hour the remaining protoplasts can be broken mechanically by stirring or centrifugation. Protease activity can be minimized by keeping the temperature low. Assuming that 95% of the protoplasts (and none of the sturdier mitochondria) are broken by stirring, the final protein concentration is 7.3 g/l, of which 2.5% is cytoplasmic enzyme. Almost all of the mitochondria (95.6%) are released during the second lysis and protoplast rupture, but they remain whole because the buffer osmolality is kept above 0.3 Os/L. Centrifugation of the final mixture produces 4.77 g/l of a pellet, of which 2.66 g or 56% is mitochondria and 0.89 g or 19%, protoplasts.

These simulations suggest an additional test for the structured model: that is, to compare its predictions to data on release of site-linked enzymes in yeast. Tests for cytoplasmic and mitochondrial enzyme release will be aided by preparation of a low-protease lytic system.

Conclusions

The simple model is conceptually straightforward and gives an approximate fit to the data over the entire range of variables studied: yeast concentration, enzyme concentration and time. The product distribution depends on the relative rates of lysis and proteolysis. Using a single enzyme preparation, as was done here, the relative rates change only with yeast concentration. In practice, however, inhibition of the protease activity, supplementation of the lytic activity with purified glucanases or mixing of lytic systems from different sources can bring about large changes in the activity ratio, which may be incorporated into the simple model by adjusting K_{mp} and k_r.

The structured model is consistent with features of lytic enzyme action and yeast structure reported in the literature. The sequential removal of the two wall layers, followed by protoplast rupture, accurately describes the early lag in protein and carbohydrate release. The presence of residual solids at long reaction times was accounted for stabilization of protoplasts by substances released from lysed cells. The structured model can be used to estimate the effects of several process alternatives, as shown in a simulation of a process for recovery of site-linked enzymes from yeast.

Acknowledgments

This work was supported by a grant from the National Science Foundation, (NSF) to whom thanks are due. One of the authors (JBH) was supported by graduate fellowships from NSF and the Josephine de Kármán Foundation during part of this work. This support is also gratefully acknowledged.

Literature Cited

1. Phaff, H.J., "Enzymatic yeast cell wall degradation", in Food Proteins: Improvement through Chemical and Enzymatic Modification, eds. R.J. Feeney and J.R. Whitaker, Adv. Chem. Series, Vol.60, American Chemical Society, 1977.
2. Asenjo, J.A., "Process for the production of yeast-lytic enzymes and the disruption of whole yeast cells", in Advances in Biotechnology, Vol III: Fermentation Products, C. Vezina & K. Singh, ed's., Pergamon, 1981; p. 295.
3. Le Corre, S., B.A. Andrews, and J.A. Asenjo, Enzyme Microb. Technol., 1985, 7, 73-77.
4. Kobayashi, R., T. Miwa, S. Yamamoto, S. Nagasaki, Eur. J. Appl. Microbiol.Biotechnol., 1982, 15, 14-19.
5. Okagbue, R.N. and M.J. Lewis, Biotech. Lett., 1983, 5, 731-736.
6. Jamas, S., C.-K. Rha, A.J. Sinskey, "Directed biosynthesis of yeast glucans with known structure-function properties", paper presented at the ACS 190th Annual Meeting, Chicago, IL, September 9-13, 1985.

7. Brake, A.J., J.P. Merryweather, D. Coit, U. Heberlein, F.R. Masiarz, G.T. Mullenbach, P.Valenzuela, and P.J. Barr, Proc. Natl. Acad. Sci. USA, 1984, 81, 4642-46.
8. Schoner, R.G., L.F. Ellis, B.E. Schoner, Bio/Technology, 1985, 3, 151-54.
9. Valenzuela, P., D. Coit, M.A. Medina-Selby, C.H. Kuo, G. Van Nest, R.L. Burke, P.Bull, M.S. Urdea, P.V. Graves, Bio/Technology, 1985, 3, 323-326.
10. D'Souza, S.F., Biotechnol. Bioeng., 1983, 25, 1661-1664.
11. Fish, N.M. and M.D. Lilly, Bio/Technology, 1984, 2, 623-627.
12. Phaff, H.J., "Structure and biosynthesis of the yeast cell envelope", in The Yeasts, Vol II, ed. R. Harrison and A. Rose, Academic Press, 1971.
13. Follows, M., P.J. Hetherington, P.Dunnill, M.D. Lilly, Biotechnol. Bioeng, 1971, 13, 549-560.
14. Marffy, F. and M.-R. Kula, Biotechnol. Bioeng., 1974, 16, 623-634.
15. Shetty, J.and J.Kinsella, Biotechnol. Bioeng., 1978, 20, 755-766
16. Ballou, C.E., "Yeast cell wall and cell surface", in The molecular biology of the yeast Saccharomyces: Metabolism and gene expression, ed. J.N. Stratton, E.W. Jones, J.R. Broach, Cold Spring Harbor, 1982 ; pp. 335-361.
17. Ballou, C.E., Adv. Microb. Physiol, 1976, 14, 93.
18. Kitamura, K., Agric. Biol. Chem., 1982, 46, 963-969.
19. Kitamura, K., Agric. Biol. Chem., 1982, 46, 2093-99.
20. Scott, J.H. and R. Schekman, J. Bacteriol., 1980, 142, 414-423.
21. Rombouts, F.M. and H. Phaff, Eur. J. Biochem., 1976, 63, 121-130.
22. Ishida, K., M. Kawai, N. Mukai, U.S. Patent 3, 809,780, May 7, 1974.
23. Asenjo, J.A., and P.Dunnill, Biotechnol. Bioeng., 1981, 23, 1045.
24. Vrsanska, M., P. Biely and Z. Kratky, Z. Allg. Mikrobiol., 1977, 17, 465-480.
25. Jeffries, T.W., and Macmillan, J.D., Carbohydr.Res., 1981, 95, 87-100.
26. Obata, T., H. Iwata, and Y. Namba, Agric. Bio. Chem., 1977, 41, 2387-2394.
27. Andrews, B.A. and J.A. Asenjo, "Continuous Culture Studies of the Synthesis and Regulation of Extracellular $\beta(1-3)$ Glucanase and Protease Enzymes in Oerskovia xanthineolytica", Biotechnol. Bioeng., 1986. (submitted for publication).
28. Hunter, J.B. and Asenjo, J.A., "Kinetics of enzymatic lysis and disruption of yeast cells: 1. Evaluation of two lytic systems with different properties." Biotechnol. Bioeng., 1986 (submitted for publication).
29. Jeffries, T.W., Ph.D. Thesis, Rutgers University, 1976.
30. Hunter, J.B. and Asenjo, J.A., "Kinetics of enzymatic lysis and disruption of yeast cells: 2. A simple model of lysis kinetics" Biotechnol. Bioeng. 1986 (submitted for publication)
31. Bernath, V.R., and W.R. Vieth, Biotechnol. Bioeng., 1972, 14, 737-745.
32. Chipman, D. Biochemistry, 1971, 10, 1714-22.
33. Asenjo, J.A., Biotechnol. Bioeng., 1983, 25, 90.

34. Lee. Y.-H., and L.T. Fan, Biotechnol. Bioeng. , 1983, 25, 939-966.
35. Hunter, J.B. and Asenjo, J.A., "A structured, mechanistic model of the kinetics of enzymatic lysis and disruption of yeast cells", Biotechnol. Bioeng., 1986 (being submitted).
36. Indge, K.J., J. Gen. Microbiology, 1968, 51, 425-432.
37. Indge, K.J., J. Gen. Microbiology, 1968, 51, 433-440.
38. Fan, L.T., Y.-H. Lee, D.H. Beardmore, Biotechnol. Bioeng.,1980, 22, 177-199.

RECEIVED March 26, 1986

3

Dual Hollow-Fiber Bioreactor for Aerobic Whole-Cell Immobilization

Ho Nam Chang[1], Bong Hyun Chung[1], and In Ho Kim[2]

[1] Department of Chemical Engineering, Korea Advanced Institute of Science and Technology, Dongdaemun, Seoul, Korea
[2] Lucky Central Research Institute, Daeduk, Korea

> Aerobically growing Escherichia coli, Aspergillus niger and Norcardia mediterranei were immobilized in the interstitial space of a dual hollow-fiber bioreactor formed by a parallel arrangement of three microporous polypropylene hollow fibers contained within a silicone tubule. All three types of cells grew well and attained high densities to reach 550–600 g dry cell weight per liter of the cell containing volume. In the cultivation of E. coli, cell growth among the fibers was not uniform and leakage of cells through the fiber walls was observed. The unlimited growth of A. niger expanded the silicone fiber and compressed the inner fibers to reduce the substrate flow rates gradually to zero. Only Nocardia mediterranei was grown successfully to make possible long term operation of 50 days or more, producing antibiotics rifamycin B with a volumetric productivity of 125 μg/mL/h based on the volume occupied by the immobilized cells. This corresponds to a 30-fold increase over the productivity of a comparable batch system.

For the past fifteen years hollow-fiber membrane bioreactors have been extensively used for immobilizing enzymes (1,2), animal cells (3), microbial cells (4,5) and plant cells (6). Immobilization of microbial cells in a hollow-fiber reactor offers some distinct advantages over other methods: cells can be easily immobilized without much preparation; primary separation of products is carried out; very little energy will be consumed in the scale-up operation. However, there are some disadvantages as well. Insoluble substrates can not be used and usually substrate pretreatment is required to prevent the fibers from being blocked. In general, polymer-based hollow fibers can not be repeatedly heat-sterilized. Inherently the transport of gas is difficult and thus the cultivation of aerobic cells with high oxygen demand becomes difficult.

Robertson and his colleagues at Stanford University have examined hollow-fiber membrane bioreactors as a means for continuous production of β-lactamase and ethanol using E. coli and S. cerevisiae immobilized in the sponge layer of asymmetric hollow fibers (7,8). In the E. coli culture cells leaked through the fiber walls, and the production of carbon dioxide was a problem in the ethanol production. Recently Robertson and Kim (9) developed a dual hollow-fiber bioreactor consisting of silicone tubules for oxygen transport and microporous polypropylene hollow fibers for substrate transport to study the production of tetracycline using Streptomyces aureofaciens. Higher productivity as compared to that of the batch fermentation was achieved. The production remained at high levels for three days and declined sharply after that for unknown reasons. The present study reports the first successful long-term cultivation of rifamycin-B producing Nocardia mediterranei and the problems encountered in growing E. coli and A. niger cells in the dual hollow-fiber bioreactor.

Materials and Methods

Materials and strains. Yeast extract, bactopeptone, malt extract and trypton were products of Difco Laboratories and glucose was from Hayashi Pure Ind. (Tokyo, Japan). E. coli (Sigma EC-1, alkaline phosphatase-rich mutant) was from Sigma Chemical Co. (St. Louis, MO) and Nocardia mediterranei (ATCC 21789) was from American Type Culture Collection. Aspergillus niger B-60 was obtained from Kubicek at Technical University Wien (Austria) who used this strain for citric acid production studies (10,11).

Bioreactor system. The reactor used in this study was constructed differently from that of Robertson and Kim (9). The reactor was a glass tubing of 30cm length (0.8 cm i.d.) in which ten dual hollow-fiber units were bundled together in a parallel assemblage. Each unit had one silicone tubule (Dow Corning, 0.147 cm i.d., 0.196 cm o.d.) that contained three microporous polypropylene hollow fibers (Enka, West Germany, 0.03 cm i.d., 0.065 cm o.d.) inside. Robertson and Kim used one polypropylene hollow fiber in which three silicone tubules were placed to make one polypropylene/silicone fiber assemblage. Thus the order in silicone and polypropylene fibers was opposite to the originally developed bioreactor. The cross section of a dual hollow fiber unit is shown in Figure 1 wherein microbial cells are supposed to grow in the restricted interstitial space between the two fiber walls. The detailed dimensions of the reactor are shown in Figure 2. The total volume of the glass tube based on the 16-cm effective length was 8.04 cm^3 and that of the interstice for cell growth was 1.12 cm^3. The cell inoculum port was covered with rubber through which inoculation could be made with a syringe needle.

Reactor operation. The polypropylene hollow fibers in the reactor were prewetted prior to inoculation with recirculation of 50% ethanol and sterilized chemically with 5% formalin solution. Then the reactor was washed by ultrafiltration of one liter of autoclaved distilled water. The reactor was placed in a water bath maintained at a desired temperature. Cells were inoculated through the inoculation port using a syringe needle. The detailed experimental setup is shown in Figure 3.

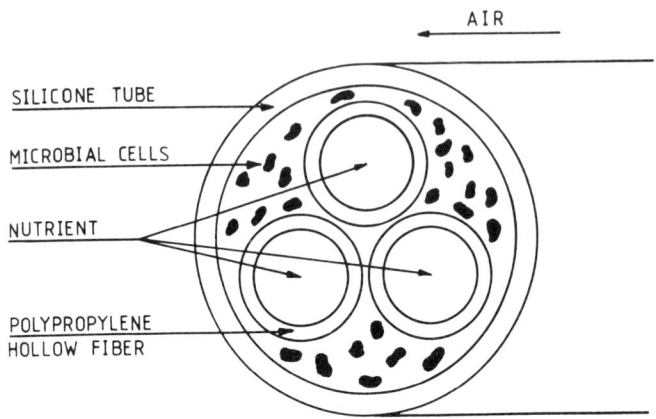

Figure 1. Cross sectional view of a dual hollow fiber unit.

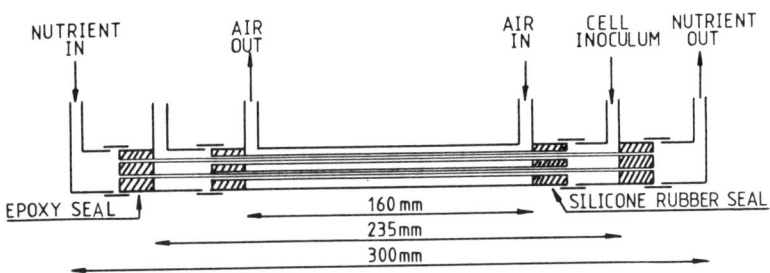

Figure 2. Detailed specification of a dual hollow-fiber bioreactor.

E. coli seed culture from the lyophilized cells was grown in a 250 mL flask with a LB medium (yeast extract, 10 g/L, trypton, 10 g/L , pH 7 adjusted with 1 N NaOH) placed on a rotary shaker (250 rpm). When the flask culture reached exponential growth phase, it was diluted 20 times and inoculated into the reactor. The medium and air flow rates were maintained at 2 mL/h and 100 mL/min, respectively. In order to see nutrient consumption during the cell growth a LB medium with 5% glucose was used. The temperature for the seed culture and the reactor operation was 37°C.

For the A. niger culture, spores grown in sugar agar slant were diluted with sterilized distilled water to a concentration of 10^8-10^9 spores/L and inoculated into the reactor. The medium for the reactor operation consisted of sucrose, 60 g/L; KH_2PO_4, 1 g/L; $MgSO_4 \cdot 7H_2O$, 0.25 g/L; NH_4NO_3, 2.5 g/L and the pH was adjusted to 3.1 with $2N$ HCl. The flow rates for the medium and air were the same as in the E. coli case. The temperature was maintained at 30°C.

The medium for N. mediterranei seed culture were: glucose, 20 g/L; yeast extract, 5 g/L; bactopeptone, 5 g/L; malt extract, 5 g/L (pH 7.3 adjusted with 1 N NaOH). The seed culture was carried out as in the E. coli culture except the temperature (30°C). When the glucose level dropped to 9 - 11 g/L, the seed culture was diluted 10 times and used in the inoculation. For comparison, a batch culture was performed with the following medium composition: glucose, 110 g/L; yeast extract, 10 g/L; bactopeptone, 10 g/L; sodium barbital, 0.7 g/L. The batch fermentation was carried out in a 500 mL flask at 30°C and at 250 rpm on a rotary shaker. For the hollow-fiber reactor the medium flow rate was 1.7 mL/h and the air flow rate was 100 mL/min. The medium composition for the reactor operation was: glucose, 20 g/L; yeast extract, 5 g/L; malt extract, 5 g/L; sodium barbital, 0.5 g/L.

Analytical methods. After the reactor operation the fibers were cut into 10 cm segments and dried in an oven at 90°C for 72 hours. The dry mass density was obtained by taking the difference between the dry mass of the cut fiber and that of an empty one of equivalent length. This difference corresponds to the biomass accumulated in the interstitial space between the inner and outer fibers. Glucose was determined by glucose analyzer (YSI model 23 A, Yellow Springs, OH) and rifamycin B was measured spectrophotometrically at 425 nm.

Microscopic techniques. Morphological examination of E. coli and N. mediterranei contained in the reactor was done in a Jeol transmission electron microscope (model 100CX). The sample was prepared according to the method by Robertson and Kim (9). For the picture of A. niger reactor the fiber was cut into 1 mm pieces, which were photographed with a light microscope.

Results and Discussion

Cultivation of E. coli. Figure 4 shows glucose concentration and pH histories during the course of E. coli cultivation in the reactor. After 4 days E. coli cells began to appear in the effluent, which means that the cells in the reactor leaked through the pores of the polypropylene membrane which were supposed to be smaller than the

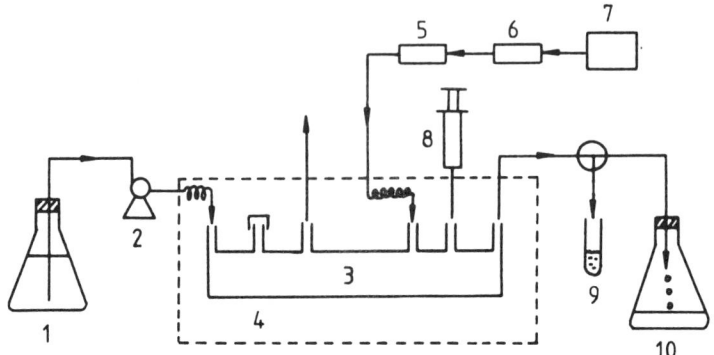

Figure 3. Schematic diagram of experimental setup: 1. medium reservoir, 2. peristaltic pump, 3. dual hollow-fiber bioreactor, 4. water bath, 5. air or pure oxygen bombe, 6. rotameter, 7. humidifier, 8. inoculum syringe, 9. sampling bottle, 10. effluent reservoir.

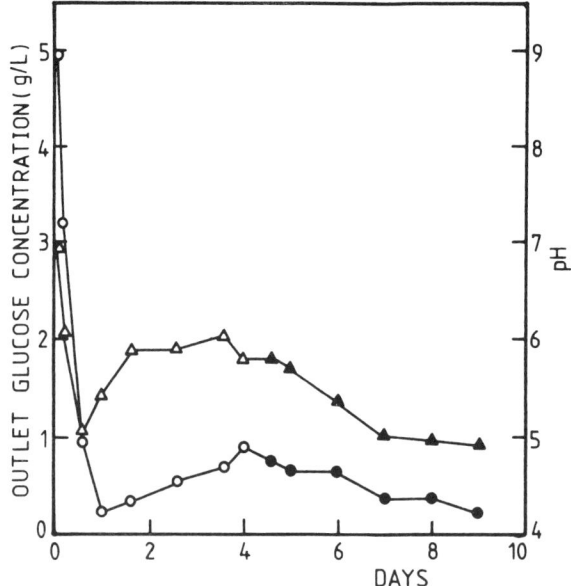

Figure 4. Glucose and pH histories in the effluent during the course of the dual hollow-fiber reactor operation. - o -, - ● - ; glucose conc., -△-, -▲-; pH. The filled circles and triangles represent the effluents containing E. coli cells.

size of E. coli cells. Figures 5(a) and 5(b) show the electron micrographs of the E. coli cells at the boundary of the polypropylene fiber and in the middle region between the two fibers. The cells were packed like tissue and some of the cells penetrated into the isotropic membrane structure. The dry cell mass was 550 g/L, which is the highest cell mass ever reported in the literature as shown in Table 1. This high cell mass compares well with 10^{12} E. coli cells/mL achieved by Inloes et al. (7) in the sponge region of a hollow-fiber reactor if we assume that the mass of a single E. coli is roughly 10^{-12}g. Leakage of cells and high cell densities seem common in this type of reactors in the case of E. coli.

After the experiment the reactor was dismantled and each of ten dual hollow-fiber units was visually examined. Only in 4 out of the ten fibers cells were densely packed, which suggests that the medium was not adequately supplied to many of these fibers. Probably the medium was not equally distributed among the fibers. In other words, in some of fibers the medium flow was not adequate to support the cell growth in the fiber. The nonuniform flow distribution among the fibers of a hollow fiber device is an intrinsic problem, which was studied in depth in the authors' laboratory (16). The work of E. coli immobilization in the dual hollow fiber reactor was reported previously from the authors' laboratory (17).

Culivation of A. niger. In the culture of A. niger B-60, the cells did not leak through the fibers. In other words, no mycelia were detected in the effluent. In all the fibers the cells grew well and appeared uniform along the fibers, but the silicone tubes were expanded and the polypropylene tubes were contracted. Figure 6(a) is the photograph of an empty silicone tube and Figure 6(b) shows the cross section of the fiber after 15 days of the A. niger growth. It was observed that the flow rate decreased gradually to zero meaning that the pumping head of a peritaltic pump was not sufficient to overcome the flow resistance exerted by the growing fungi. Thus the continuous operation of the reactor was not feasible. It is suggested that the control of the cell growth be needed after a growth period by switching the culture to a nitrogen deficient medium or by some other means.

Production of rifamycin B by N. mediterranei. Figure 7 shows the results of shake flask culture for rifamycin B production. After 8 days of fermentation 60 g/L of glucose was consumed and 820 µg/mL of rifamycin B was produced. This gives a volumetric productivity of 4.3 µg of rifamycin/mL/h. The continuous production of rifamycin B in the dual hollow-fiber bioreactor is shown in Figure 8. The successful production of the antibiotics continued more than 50 days without showing signs of decreased production. This is in contrast to the tetracycline production by Robertson and Kim that lasted for a few days. The cause of this decline in Robertson and Kim's work has not been understood. Stabilities in the antibiotics production in the dual hollow- fiber bioreactor are speculated to be associated with reactor design or producing organism or both.

Two distinct differences are noted between the flask culture and the present reactor system. First, the effluent concentration of rifamycin B was ca. 1/10th of that obtained in the flask culture.

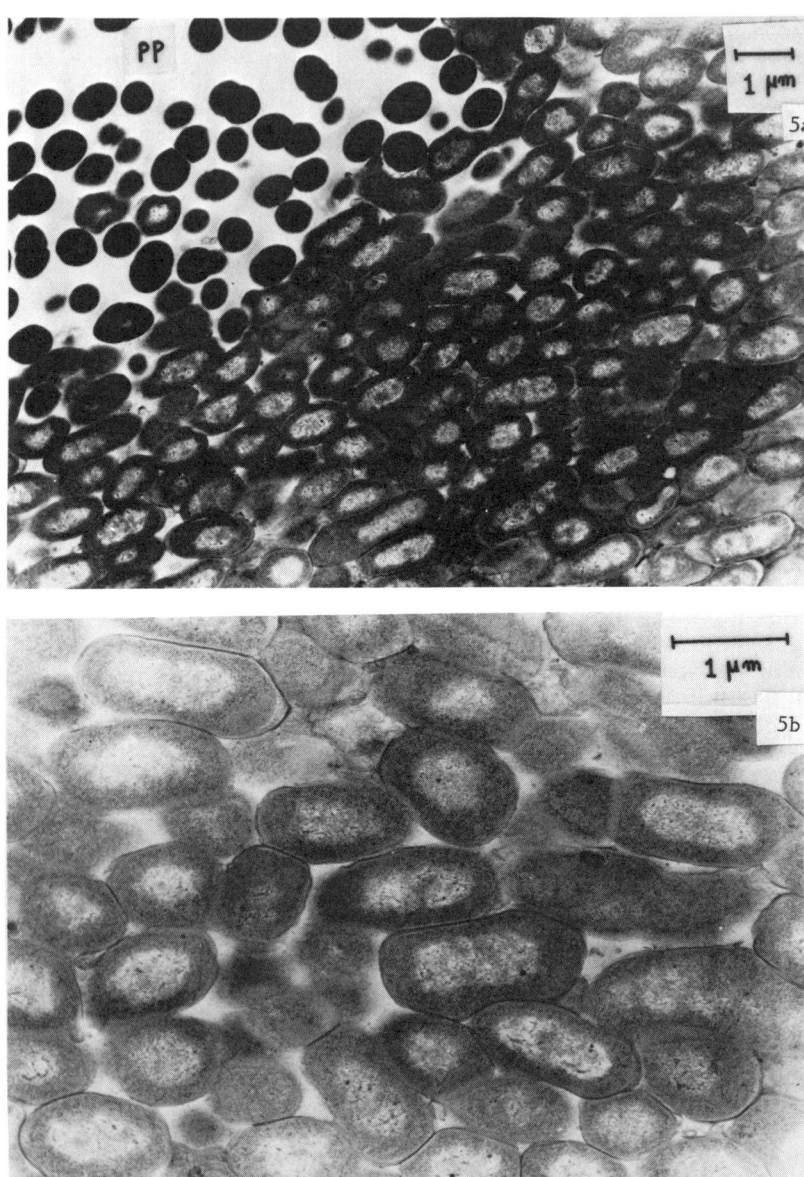

Figure 5. Electron micrographs of densely packed E. coli K-12 cells. (a). The cells at the boundary of the polypropylene fiber (pp). Magnification, 10,000X. (b). The cells in the middle space between the silicone tube and the polypropylene fiber. Maginification, 20,000X.

Table 1. Comparison of cell mass of E. coli in various fermentations.

System	Cell density	Reference
o Shake-flask culture	1 - 2 g/L	
o Submerged-culture under controlled conditions	10 g/L	(12,13)
o Submerged-culture with pure oxygen supply and semi-continuous feeding of glucose at 22°C.	55 g/L	(14)
o Immobilization with carrageenan beads	4.5×10^{10} viable cells/mL	(15)
o Immobilization in a hollow-fiber bioreactor	10^{12} cells/mL	(7)
o Immobilization in a dual hollow-fiber bioreactor	550 g/L	This work

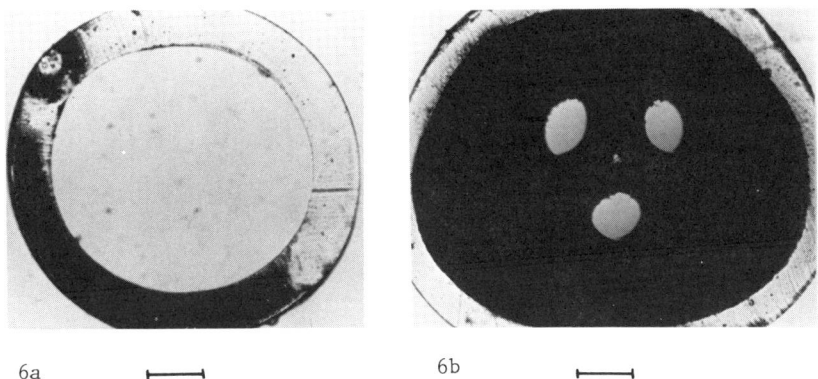

6a 6b

Figure 6. Expansion of silicone tube and contraction of polypropylene tubes by growing A. niger B-60. The length scale shown in the pictures is 250 μm. (a). Cross section of an empty silicone tube. (b). Deformed bioreactor.

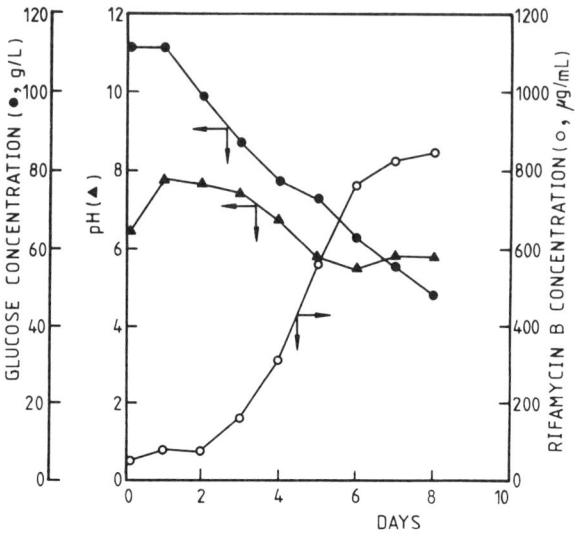

Figure 7. Rifamycin B production by shake flask fermentation.

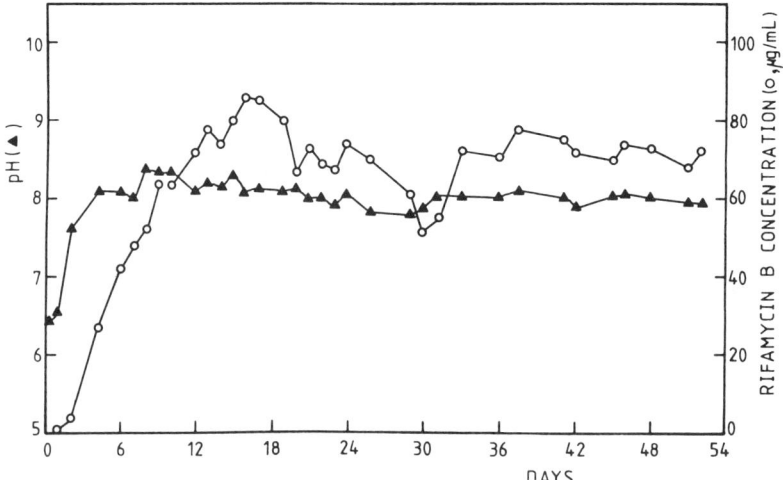

Figure 8. Continuous production of rifamycin B in a dual hollow-fiber bioreactor.

This is not small at all if we consider that the residence time in the hollow-fiber reactor was 12 minutes based on the total fiber lumen voume of 0.339 cm^3 while that in the flask culture was 8 days. The volumetric productivity of the reactor was 125 µg/mL/h based on the void voulme of the reactor where the cells were actually immobilized. This was about thirty-fold as compared to that of the flask culture. This number drops to 15 or 10 if we include the reactor volume or the volume of the glass tubing used. This high productivity comes essentially from a highly dense cell mass in the reactor shown in the electron micrograph (Fig. 9). The measured dry cell mass was 600 g/L. The cells neither penetrated into the propylene fibers nor expanded the tubes. This growth characteristics is in good contrast to that of E. coli or A. niger cells. The cells grew uniformly along the fibers, which made possible the successful long-term operation of the reactor.

The main advantages of a hollow-fiber reactor system are: very little energy will be consumed in the aeration; primary purification is accomplished concurrently with the production. In conventional antibiotics or citric acid fermentation with A. niger much energy is consumed in several days' of continuous aeration and mixing of viscous fermentation broths which adds up to a substantial portion of final production costs. If this membrane bioreactor is ever successful in a scale-up operation, there will be a tremendous savings in aeration and mixing costs. Also the benefit of primary separation can never be underestimated because simpler downstream processing is a key to production cost reduction. However, as yet, much work needs to be done for this reactor to become attractive for

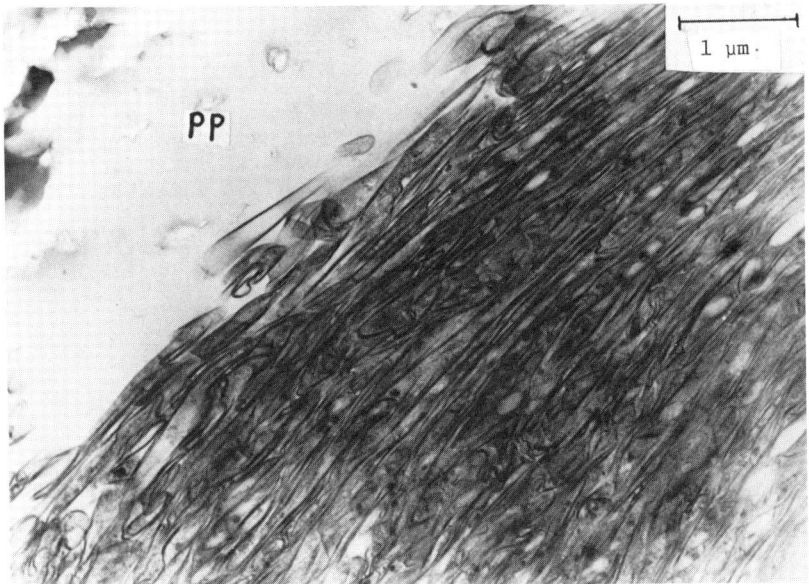

Figure 9. Electron micrograph of densely packed Nocardia mediterranei (ATCC 21789) cells near the polypropylene hollow fiber (magnification, 25,000X).

industrial production of valuable materials. Achieving higher volumetric productivity is certainly an advantage, but the final product concentration is too low for any real recovery process to be considered as compared to that in batch system. Substrate diffusion limitation through the membrane can be blamed for this low product concentration, but this is not the case considering that more than 80% of glucose is consumed during the 10 minutes' residence time in the E. coli reactor. Oxygen limitation can be a cause for this. Perhaps the most important reason is that cells in a distressed state can not function as well as the cells in suspension. The water content of the cells in the hollow-fiber would be around 40% for the cells to attain such a high dry cell weight. Currently we are working on ways of increasing the final product concentration of rifamycin comparable to that in the batch system and are trying to improve the stability in the operation of reactor for A. niger cells.

Literature Cited

1. Rony, P. R. Biotechnol. Bioeng. 1971, 13, 431.
2. Waterland, L. R.; Michaels, A. S.; Robertson, C. R. AIChE J. 1974, 20, 50.
3. Knazek, R. A.; Gullino, R. M.; Kohler, P. O.; Dedrick, R. L. Science 1972, 178, 65.
4. Kan, J. K.; Shuler, M. L. Biotechnol. Bioeng. 1978, 20, 217.
5. Vick Roy, T. B.; Blanch, H. W.; Wilke, C. R. Biotechnol. Lett. 1982, 4, 483.
6. Shuler, M. L. Ann. N.Y. Acad. Sci. 1981, 369, 65.
7. Inloes, D. S.; Smith, W. J.; Taylor, D. P.; Cohen, S. N.; Michaels, A. S.; Robertson, C. R. Biotechnol. Bioeng. 1983, 25, 2653.
8. Inloes, D. S.; Taylor, D. P.; Cohen, S. N.; Michaels, A. S.; Robertson, C. R. Appl. Environ. Microbiol. 1983, 46, 264.
9. Robertson, C. R.; Kim, I. H. Biotechnol. Bioeng. 1985, 27, 1012.
10. Habison, A.; Kubicek, C. P.; Röhr, M. FEMS Microbial Lett. 1979, 5, 39.
11. Mischak, H.; Kubicek, C. P.; Röhr, M. Biotechnol. Lett. 1984, 6, 425.
12. Elsworth, R.; Miller, G. A.; Whitaker, A. R.; Kitching, D.; Sayer, P. D. J. Appl. Chem. 1968, 17, 157.
13. Phares, E. F. In "Methods in Enzymology"; Colowick, S. P.; Kaplan, N. O., Ed; Academic Press: New York, 1971; vol. 22 p. 157.
14. Shiloach, J.; Bauer, S. Biotechnol. Bioeng. 1975, 17, 227.
15. Wada, M; Kato, J.; Chibata, I. Eur. J. Appl. Microbiol. Biotech. 1979, 8, 241.
16. Park, J. K.; Chang, H. N. AIChE J. (accepted).
17. Chung, B. H.; Chang, H. N.; Kim, I. H. Korean J. Appl. Microbiol. Biotechnol. 1985, 13, 209.

RECEIVED March 26, 1986

A Membrane Reactor for Simultaneous Production of Anaerobic Single-Cell Protein and Methane

R. K. Finn and E. Ercoli

School of Chemical Engineering, Cornell University, Ithaca, NY 14853

> Single-cell protein can be produced from agricultural residue anaerobically in yields of about 20% (wt. cells per wt. substrate) by using a mixed culture of rumen bacteria. Even higher cell yields should be possible. To achieve high cell densities, it is proposed that acidic end products be removed by cyclic microfiltration into a methanogenic fermentor. Preliminary experiments suggest that such a tandem fermentation should be feasible on a continuous basis. More data are needed for an economic evaluation.

Very little attention has been given to the possibilities for anaerobic production of single-cell protein (SCP) from cheap carbohydrate residues (1,2). The reason for dismissing any anaerobic process is that cell yields, according to classical Embden-Meyerhof catabolism, are only 10 to 15% of the substrate fermented. In contrast, aerobic cell yields of 50 to 60% are easily obtainable.

However, there are highly efficient anaerobic cells. These include the acetogens, propionic bacteria, and above all the various rumen bacteria. The latter can attain cell yields on carbohydrate of 30 to 35 dry weight (3,4).[1] Such yield values are already corrected for any polysaccharide formation, and in fact the protein content of rumen bacteria is about 60% (2). We therefore have been considering their use as a protein feed supplement for monogastric animals like chickens or pigs.

The ruminant animal and rumen microorganisms exist in a reciprocally beneficial relationship, in which cellulose and other plant carbohydrates are fermented by the rumen microbes to form chiefly CO_2 and volatile fatty acids (VFA). The microorganisms are adapted to live between pH 5.5 and 7.0, in the absence of oxygen,

---

[1]These high yields result from a combination of factors including low maintenance energy, higher than normal cell yields per mol of ATP, and finally excess ATP production, which can involve "anaerobic respiration" with cytochrome b in a fumarate cycle.

at temperatures of 39° to 40°C, and in the presence of moderate concentrations of fermentation products. These volatile fatty acids, chiefly acetic and propionic, are absorbed through the rumen wall to provide energy for the animal. Removal of the acids is essential because at concentration above about 0.3% they do inhibit cell growth. Within the rumen though, microorganisms grow very efficiently, and thereby provide single-cell protein for their host animals. Our challenge as engineers is to duplicate in vitro such performance.

Some of the basic ideas of the anaerobic SCP process we are developing at Cornell are summarized below.

Anaerobic SCP Process

1) Rapid growth of mixed rumen bacteria (e.g. μ = 0.16 hr^{-1} on starch) to high cell densities at cell yields of 30-35%
2) Membrane removal of inhibitory acid products
3) The acids feed slower growing methanogenic bacteria in a separate fermentor
4) The CH_4 generated thus can be used to dry the SCP product
5) Rapid interchange between the two fermentors is possible with alternating pulsed microfiltration.

The key to economic cell production is rapid growth to cell densities like those in the rumen, namely 10^{10} or 10^{11} cells/ml. Acidic end-products are used to feed a methane generator, so that most of the carbon is recovered in a useful form. An unusual feature of this process is that rapid ultrafiltration rather than slow dialysis can be used to feed the methane fermentor. Insoluble substrates such as starch, hemicellulose or cellulose are retained within the rumen fermentor by appropriate membranes. The rapid interchange of soluble acids between the two fermentors allows only a low steady-state concentration to develop in the rumen fermentor because conversion to methane proceeds simultaneously in the second fermentor.

Additional features of the process are listed below.

1) Broad range of insoluble carbohydrates fermented...crude mixed cultures of defined consortia of rumen bacteria.
2) Sterile operation may be unnecessary because rumen conditions select for a very specialized mixed population.
3) Safety and nutritional value of the SCP product for use as an animal feed has already been proven by studies on ruminant animals.
4) A lower productivity than in an aerobic process is to be expected because of a somewhat lower cell yield. Overall costs may still be competitive because of simpler equipment and energy savings.
5) Fed-batch or continuous operation seems feasible.

We have done only preliminary experimental work using mainly glucose and sugar beet pulp as substrates. We have not yet combined the methanogenic step; instead we have used a buffered salts medium in the second chamber to remove inhibitory end products.

Procedures. The basal medium contained mineral salts, yeast extract (0.1 g/ℓ), trypticase (0.1 g/ℓ), cysteine hydrochloride (0.4 g/ℓ) as a reducing agent, and reazurin as a redox indicator. The base used to maintain a constant pH was sodium carbonate. Some media included hemin (2 to 6 mg/ℓ) because most strains of Bacteroides ruminicola, a major type of rumen bacteria, are stimulated by the addition of small amounts of hemin to the medium (5). For example, the molar growth yield of Bacteroides fragiles subsp fragilis increased from 17.9 to 47.0 (g dry weight cell per mol of glucose) when 2 mg/ℓ of hemin were added (6). Media were inoculated with fresh rumen fluid taken from a cow fed with grain and hay. The samples were used within two hours after removal.

Protein was determined by the method of Lowry (7) after hydrolysis with 0.2N NaOH (100°C, 15 min). Total nitrogen was measured by the micro-Kjeldahl method with sulfuric acid/hydrogen peroxide reagent; the ammonia was detected with Nessler's reagent. Glucose was measured by standard colorimetric assay using dinitrosalicylic acid. Starch was hydrolyzed with concentrated HCl and then determined as sugar.

Results. To establish optimum growth conditions, we used in the early experiments a low concentration of the carbon source. Removal of the inhibitory acids is then unnecessary.

Table I shows results for a mixed population of rumen bacteria. The fermentations were complete (essentially no residual glucose) after 6 to 7 hours.

Table I. Growth of Rumen Bacteria on Glucose

Substrate Conc. (g/ℓ)	Hemin (mg/ℓ)	Growth rate (h^{-1})	Nitrogen* Fixed (g/ℓ)	Est'd cell** yield (g/g substr.)
5	–	0.61	0.071	0.15
5	6	0.66	0.090	0.19 (0.18)
15	–	0.60	0.126	0.08

*Net utilization of soluble nitrogen from the medium, i.e. converted into biomass.
**The cell yield was estimated from the nitrogen fixed, assuming 50% protein content for the cells, 15% nucleic acids. The value shown in parenthesis was measured directly by dry weight.

The results in Table I show that cell yields are lower than expected but that added hemin stimulates growth. The low yield of cells at higher glucose concentration may be caused by accumulated acids suppressing growth while allowing fermentation to proceed.

Using starch as substrate yields were also low. An explanation may be that Streptococcus bovis, a homofermentative lactic acid organism common in the rumen grows rapidly on starch with low yield of biomass (8). Table II shows some of the results for fermentation of sugar beet pulp (SBP).

Table II. Growth of Rumen Bacteria on Sugar Beet Pulp, (5% inoculum)

Substrate Conc. (g/ℓ)	Hemin (mg/ℓ)	Nitrogen fixed (g/ℓ)	Est'd cell yield (g/g subst. supplied)
5	-	0.067	0.14
5	2	0.075	0.16
5	4	0.074	0.16
5	6	0.086	0.18
10	6	0.123	0.12
20	-	0.203	0.10

The yield values in Table II are based on substrate supplied; the quantity unfermented could not be readily measured. The incorporation of nitrogen from the medium is similar to that observed for glucose.

Growth rates on sugar beet pulp were slower than on glucose, as indicated by comparison of the solid lines in Figure 1, where nitrogen in the solids fraction provides a measure of biomass. The rate of acid production (broken lines in Figure 1) was also slower for the sugar beet pulp; from the slopes of such lines one can estimate the rate at which it will be necessary to remove acids from a fermentation with higher substrate concentrations.

Early experiments were done using simple dialysis in a 2-chambered membrane apparatus constructed from 3-inch glass pipe (Figure 2). Starch was added periodically to the rumen bacteria so as to simulate a fed-batch operation. A dialysis membrane was held between flanges separating the two chambers. The rate of acid removal was too slow, however, in this simple dialysis apparatus. Consequently, a microfiltration chamber was arranged as shown in the diagram of Figure 3. Plugging or fouling of the microfiltration membrane can be avoided by changing the pressure periodically on either side of the membrane. Flow then oscillates across the membrane. Such cycling can be accomplished without changing the pressure in either of the fermentors by using two small centrifugal pumps and the valve arrangement shown at the top of the diagram. Both pumps operated continuously. The solenoid valves were alternately opened and closed on 30-second repeat cycles. Back-pressure valves were set to provide circulation through the membrane chambers while maintaining also a positive pressure for microfiltration. Each of the fermentation chambers had a working volume of 1.5 liters and there was an additional holdup of 0.5 liters on each side of the filtration chamber. A single membrane of polytetrafluoroethylene was used. It had a pore size of 0.22 micrometers and an area of 155 cm . Preliminary experiments have indicated

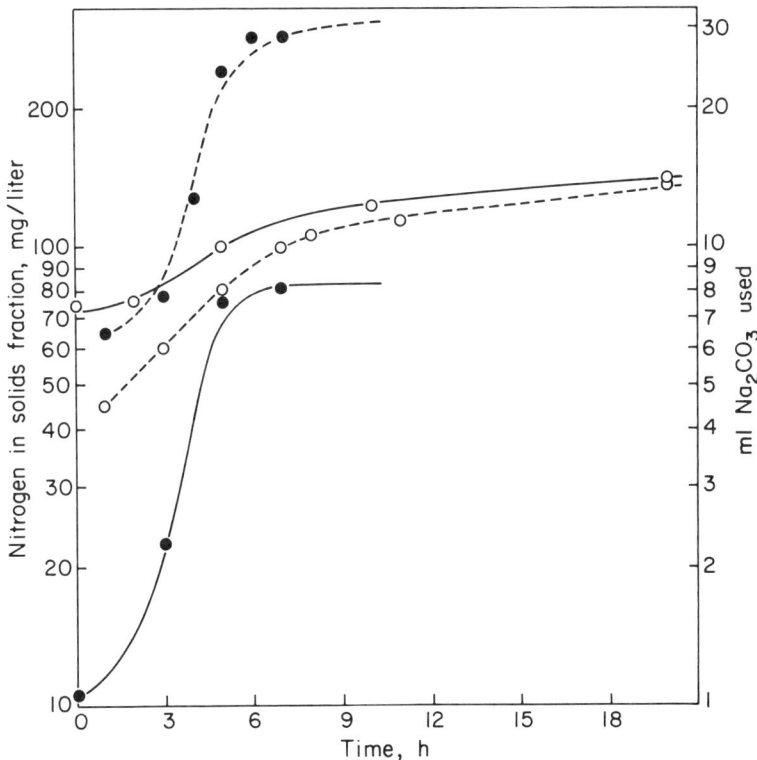

Figure 1. Time course of the rumen fermentation. Solid lines show nitrogen in the solids fraction for 5 g/liter glucose (solid circles) and for 5 g/liter sugar beet pulp (open circles). The high initial value for sugar beet pulp represents its protein content as received.

Broken lines show the corresponding amounts of 25% aqueous Na_2CO_3 used for maintaining constant pH during fermentation of the glucose or sugar beet pulp.

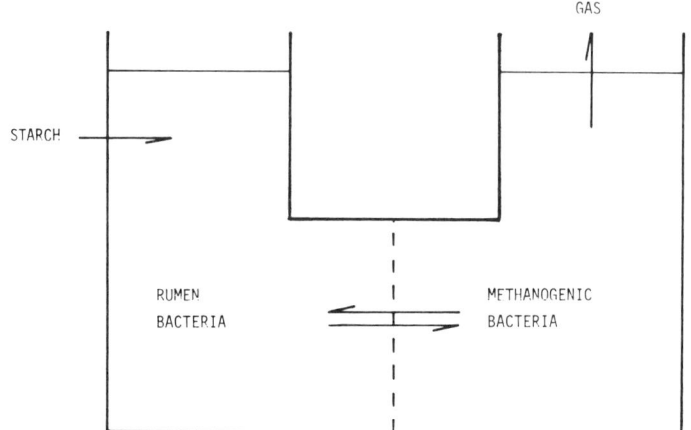

Figure 2. Diagram of apparatus using two 3-inch glass elbows with a dialysis membrane at the flange separating the two chambers. No forced flow through the membrane.

1 BACK-PRESSURE VALVES
2 SOLENOID VALVES

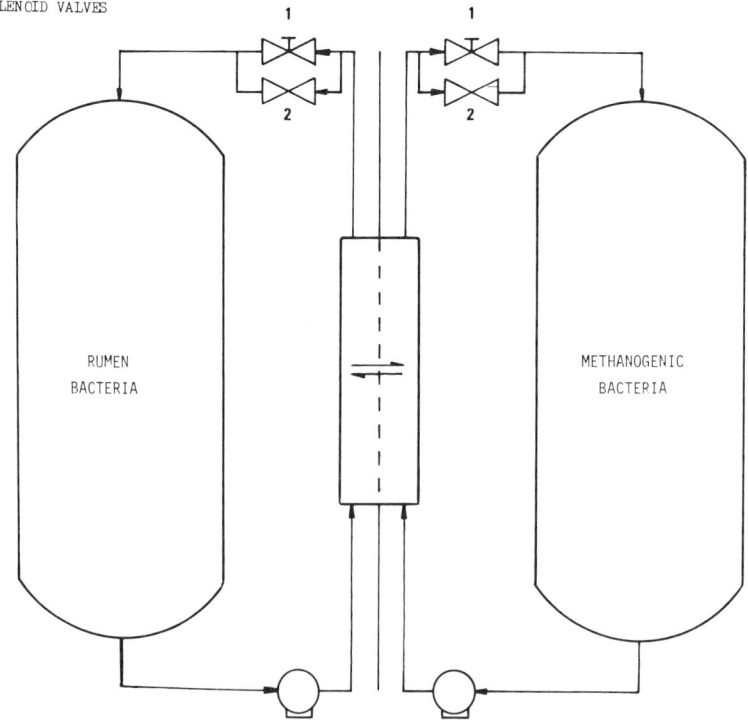

Figure 3. Process scheme for microfiltration with cycling flow back and forth between the two fermentation chambers.

that it is possible to get a net interchange of the total volume ten times per hour, even with rumen bacteria and starch particles present. Longer cycle times than 30 seconds are needed, however, to allow for adequate mixing of the permeate into the bulk fluid.

From the slope of the curve for addition of sodium carbonate to the rumen fermentation of starch, one can estimate tht the net rate of acid formation is equivalent to 36 millimoles of acetic acid per liter per hour. If a level of no more than 0.2% acid must be maintained in the rumen fermentor, then the exchange rate across the membrane must be at least one liter per hour or half of the total reactor volume each hour. This is well below the observed exchange rate of 20 liters per hour.

In summary, we have described an anaerobic process for single-cell protein from crude carbohydrates. The inhibitory by-products are simultaneously converted into methane. Mass transfer limitations can be avoided by using microfiltration rather than dialysis. Further study of the kinetics and improvements in yield will be necessary in order to make an economic comparison with other processes for single-cell protein.

Acknowledgments

We thank Professor James B. Russell, Department of Animal Science at Cornell, for many helpful discussions and acknowledge support of one of us (E.E.) from Consejo Nacional de Investigaciones Cientificas y Tecnicas de la Republica Argentina.

Literature Cited

1. Rolz, C. and Humphrey, A., "Microbial Biomass from Renewables: Review of Alternatives," Adv. Biochem. Eng. $\underline{21}$, 1-53 (1982).
2. Mehta, K.I. and Callihan, C.D., "Production of Protein and Fatty Acids in the Anaerobic Fermentation of Molasses by $\underline{E. ruminantium}$," J. Am. Oil Chemists Soc. $\underline{61}$, 1728-1734 (1984).
3. Isaccson, H.R., Hinds, C., Bryant, M.P., and Owens, F.N., "Efficiency of Energy Utilization by Mixed Rumen Bacteria in Continuous Culture," J. Dairy Sci. $\underline{58}$, 1645-1659 (1975).
4. Russell, J.B. and Baldwin, R.L., "Comparison of Maintenance Energy Expenditures and Growth Yields Among Several Rumen Bacteria Grown in Continuous Culture," Appl. Environ. Microbiol. $\underline{37}$, 537-543 (1979).
5. McCall, D. and Caldwell, D., "Tetrapyrrole Utilization by Bacteroides ruminicola, J. Bacteriol. $\underline{131}$, 809-814 (1977).
6. Macy, J., Probst, I., and Gottschalk, G., "Evidence for Cytochrome Involvement in Fumarate Reduction and ATP Synthesis by Bacteroides fragilis in the Presence of Hemin," J. Bacteriol. $\underline{123}$, 436-442 (1975).

RECEIVED March 13, 1986

SEPARATION AND CONCENTRATION PROCESSES

5

Membrane Processes in the Separation, Purification, and Concentration of Bioactive Compounds from Fermentation Broths

Enrico Drioli

Instituto di Principi di Ingeneria Chimica, Università degli Studi di Napoli, Piazzale Tecchio, 80125 Napoli, Italy

```
          The potential of membrane separation techni-
          ques (such as cross-flow microfiltration(MF),
          ultrafiltration (UF), Reverse Osmosis (RO)and
          electrodialysis (ED) ) and membrane reactors
          in the treatment of fermentation broths are
          huge. The synergistic effects obtainable by
          designing the overall biotechnological pro-
          cess combining various membrane technique are
          particularly significant.
          In this paper experimental results are descri-
          bed which refer to processes of industrial
          interest studied assuming membrane technolo-
          gies as the best available.
          The separation,purification and concentration
          of a thermosensitive bioactive compound from
          a lysate has been carried out combining UF,
          ion exchange and RO with significant cost re-
          duction and productivity increase. Enzyme mem-
          brane reactors have been used for triglyceride
          enzymatic hydrolysis and product separation.
          Thermophilic,thermostable enzyme ultrafiltra-
          tion membrane have been prepared, and used in
          high temperature lactose hydrolysis.
```

The term Downstream Processing in Biotechnology refers to the chain of unit operations that are combined into a system for the recovery, purification, separation and concentration of the products at the lowest possible cost and highest possible recovery factor and quality.
The recovery step generally represents a large part of the

0097-6156/86/0314-0052$06.00/0
© 1986 American Chemical Society

overall capital investiment in a fermentation plant and
its cost efficiency is a key factor for the production of
biotechnological compounds.
The recovery of bioactive materials from the fermentation
broth is generally complicated by the fact that the bio-
products are in very low concentration in these often unstable,non-newtonian systems. The downstream processing is
a key area for further development of biotechnology. Membrane technologies and particularly UF,MF and RO can be
considered as broad core technologies in this industrial
segment (1).
In Table I are summarized the market values of various products of interest in biotechnological processes. Their
costs of production on large scale will be affected significantly and positively by using membrane technology (2).
In Table II are summarized the various membrane processes
of interest for biotechnology.
Those systems can contribute in general to : improvements
in methods for recovery and reuse of enzymes or whole
cells; development in methods for biocatalyst immobilization; development of enzyme membrane reactors.

Membrane Systems in Downstream Processes

Traditional fermentation generally involves many process
steps which are inefficient and uneconomical in terms of
utilization of raw materials,recovery of product, and energy consumption. Continuous cross-flow membrane micro-fil-
tration and ultrafiltration, when correctly introduced,
have been shown to improve this process significantly.
The possibility of treating liquids containing suspended
solids broadens the scope of membrane filtration. Commer-
cially available capillary membranes and tubular membranes
have been used for the concentration of broths of bacteria,
yeasts and moulds, and for sterile filtration of solutions
of proteins, enzymes, vaccines and amino acids (3).
Fouling problems have been solved by a periodic backflush-
ing of the membrane.
Fluids containing upwards of 50% to 60% suspended solids
by volume can be pumped through a well-designed membrane
module easily achieving very high liquid-phase recoveries
from whole fermentation broths (2). When ultrafiltration
membranes are used on the broths (or on the permeate from
a microfiltration stage), retention of the solubilized macromolecular solutes, as well as particulate and colloidal
material, will be accomplished, giving a filtrate containing only relatively low molecular weight solutes.

Table I. Total market values for the various product categories.

Product category	Number of compounds	Current value ($ millions)
Amino acids	9	$ 1,703.0
Vitamins	6	667.7
Enzymes	11	217.7
Steroid hormones	6	376.8
Peptide hormones	9	263.7
Viral antigens	9	N|A
Short peptides	2	4.4
Nucleotides	2	72.0
Miscellaneous proteins	2[a]	300.0
Antibiotics	4[b]	4,240.0
Gene preparations	3	N|A
Pesticides	2[b]	100.0
Aliphatics :		
Methane	1	12,572.0
Other	24[c]	2,737.5
Aromatics	10[c]	1,250.9
Inorganics	2	2,681.0
Mineral leaching	5	N|A
Biodegradation	N|A	N|A
Totals	107	$ 27,186.7[d]

[a] Only two of a number of compounds are considered here.
[b] These numbers refer to major classes of compounds; not actual numbers of compounds.
[c] These numbers refer only to those compounds representing the largest market volume in classes specified in the thext.
[d] Current value excluding methane = $14,614,700.000

SOURCE : Genex Corp. (1).

Table II. Membrane Processes Used Today in Biotechnology

Process	Membrane Type	Driving Force	Mass Separation Mechanism	Area of Application
Microfiltration	Symmetric microporous polymer membrane. Pore size 0.05-10 μm	Hydrostatic pressure 1-5 bar	Sieving mechanism, pore size and particle diameter determine separation characteristics	Sterile filtration, clarification, cell harvesting bacteria, viruses separation.
Ultrafiltration	Asymmetric microporous polymer membrane. Pore size 1-50 nm	Hydrostatic pressure 2-10 bar	Sieving mechanism, pore size, and particle diameter determine separation characteristics	Separation, concentration and purification of macromolecular solutions such as proteins, enzymes, polypeptides, etc.
Reverse Osmosis	Asymmetric membrane with homogeneous skin and microporous sub- or support structure	Hydrostatic pressure 10-100 bar	Solution-diffusion mechanism, solubility, and diffusivity of individual components in the homogeneous polymer matrix determine separation characteristics.	Concentration of microsolutes, such as salts, sugars, amino acids, etc., recovery of water from microbiological processes.
Membrane Distillation	Symmetric or asymmetric mainly hydrophobic microporous membrane	Partial vapor pressure gradient introduced by a temperature difference	Partial vapor pressure, separation mechanism is the same as in distillation.	Separation volatile organic solvents such as acetone, ethanol, etc. from aqueous fermentation solution.
Pervaporation	Asymmetric membrane with homogeneous skin and microporous substructure.	Partial vapor pressure gradient 0.001 to 1 bar	Solution-diffusion mechanism, solubility and diffusivity of individual components in the polymer matrix determine separation characteristics.	Separation of organic solutions such as ethanol, butanol, acetic acid, etc. from aqueous solutions, especially separation of azeotropic mixtures.
Electrodialysis	Cation- and anion-exchange membrane	Electrical potential difference	Electric charges of particle	Removing salts, acids, and bases from fermentation broths, separation of amino acids, etc.
Liquid supported membranes	Symmetric or asymmetric microporous membranes supporting liquid phase	Chemical potential gradients	p.e. Carrier transport	Selective removing of salts, bioactive compounds, etc.

This filtrate might be also more easily treated by reverse osmosis if its concentration of low molecular weight compounds if of interest.
Those processes can be considered ideal pretreatment for reverse osmosis processes. The concentration and purification of antibiotics by sequential UF and RO,is an example. The removal of antigenic contaminants present in biological mixtures via the combined use of immunocomplexation and ultrafiltration has also been suggested(3).
An interesting application has been developed at industrial scale for the recovery of a particularly unstable amino acid from a lysate (4). The traditional downstream process after yeast enrichment in a fermentor consisted of a) centrifuge for cells separation; b) thermal lysis; c) centrifuge for solid part separation; d) chemical precipitation using specific complexing agents such as picric acid; e) centrifugation; f) purification and transformation into a compound soluble in water; g) precipitations with an organic solvent; h) filtration; i) purification on activated carbon; l) filtration; m) lyophilization (see Figure 1). The study carried out on this process showed the possibility of decreasing significantly the production costs,and time of operation, increasing product quality and process safety by using ultrafiltration in place of purification by precipitation with organic solvents (e.g. methanol).
A recovery factor of 97-98% of S-adenosyl-L-methionine was obtained, using ultrafiltration capillary membranes (Berghof FDR) with cut-off of the order of 10.000 M.W. at 1 atm applied pressure and 20 °C. Fluxes of the order of 30 $l/m^2 h$ were normal at steady state.
Pyrogens were completely absent in the permeate product.
At HPLC the product purity was particularly high. The possibility of combining UF with ion-exchange columns and final concentration of the eluate by reverse osmosis has been also analyzed in detail (see Figure 2).
Reduction of the costs for column regeneration, and waste water treatment was observed. High recovery factor up to 90% was obtained in the RO concentration step, with final concentrations higher than 80 g/l.
The amount of activated carbon used in the final purification step decreased moreover from 1 Kg to 150 g. per Kg of raw material treated. The fact that no organic solvents were required in all the process gave an overall cost reduction of few hundred dollars per Kg of final bioactive product.

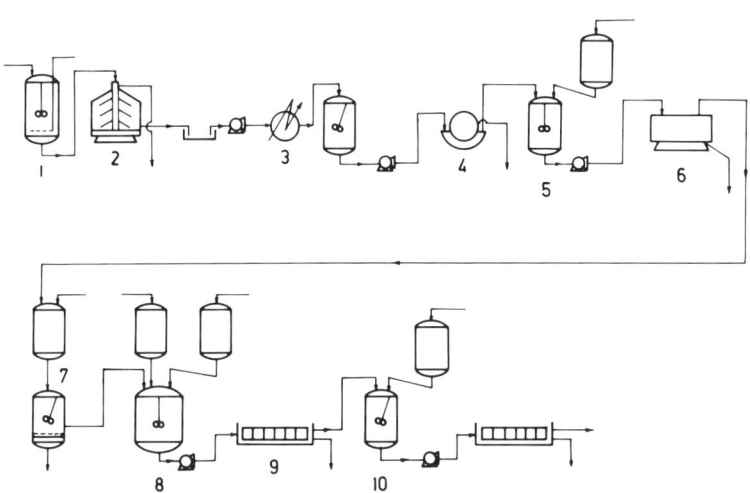

Figure 1. Traditional downstream process: 1) enrichment;
2) centrifugation; 3) thermal lysis; 4) vacuum filter;
5) chemical precipitation; 6) centrifugation; 7) purification by solubilization; 8) reprecipitation; 9) filtration;
10) active carbon; 11) filtration; 11) lyophilization.

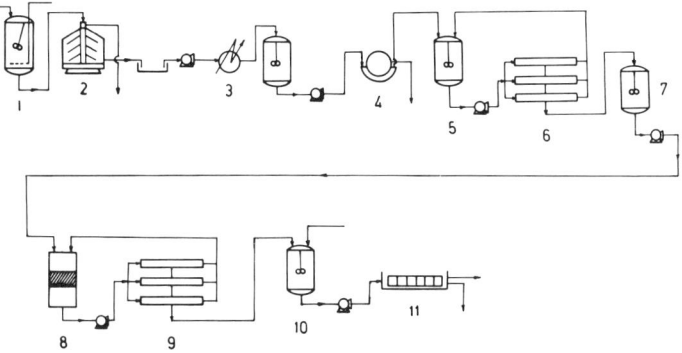

Figure 2. Modified Process : 6) ultrafiltration; 8) ion exchange; 9) Reverse Osmosis.

Enzyme Membrane Reactors

In the previous examples the membranes have been considered generally as semipermeable barriers for the separation of small molecules from bigger ones. When in parallel to the separation a chemical reaction takes place in the bulk solution or in the membrane itself, the system may be identified as a true membrane reactor. A classical example is a stirred-tank enzymatic reactor connected by a continuous recirculation loop to an ultrafiltration or dialysis unit. Such a system, when well designed, permits the continuous removal of the reaction products from the bulk solution without loss of enzyme (or the insoluble or macromolecular substrate).

The use of membrane separation as a component of continuous fermentation systems is growing in interest. In this case it is possible to increase the productivity of product-inhibited fermentation, for example, by the continuous removal of the low molecular weight products. The enzymatic degradation of cellulose to alcohol, where the microorganisms used are inhibited at an alcohol concentration higher than 12%, might be improved by the use of this concept.

Recently, a similar process has been applied by Degussa for the production of L-amino acids. In this case, L-amino acids are obtained by biocatalytic division of synthetically-produced acetyl DL-amino acids by means of enzymes. Unlike the previous type of fixed-bed reactor with carrier-located acylase, the new approach employs the enzyme in soluble form, and uses a membrane for separating the enzyme from the reagent solution. This avoids losses at the immobilizing stage and reduces enzyme consumption (5).

Other advantages are that the enzyme content can be continually adjusted and the product solution is obtained free from pyrogens. The continuous removal of toxic products might allow significant increases in cell mass and total product yield in batch fermentation for growth of bacteria. The possibility to apply those concepts to the enzymatic hydrolysis of triglyceride is under investigation in our laboratory (6). Traditional chemical hydrolysis requiring operations at high temperature and pressure make the process non-competitive. The enzymatic hydrolysis can be carried out at room temperature and atmospheric pressure. The possibility to prepare and to separate glycerine and acids in the same step appears of particular interest. In our study natural olive oil was used as raw material. An enzymatic membrane reactor was used based on ultrafiltration capillary membranes. The enzyme (lipase from Candida cylindracea,2975 U/mg) was dynamically immobilized in a gel form on the internal membrane surface. Increase of enzyme stability and separation of the reaction products were obtained. A continuous recirculation of the oil substrate in the capillary membranes was carried out at axial flow rate and applied pressure useful to mantein an high enzyme concentration at the membrane-solution interface (Figure 3). In Figure 4 some typical experimental results are presented which show the degree of conversion of the triglycerides as function of time. Only glycerine was present in the permeate. Microporous hydrophobic capillary membranes can also be used in this study to distribute small oil droplets in the water phase where the enzyme is dissolved. This emulsion can be controlled in droplets size and formation rate by changing the transmembrane pressure and axial flow rate.

Enzyme Membranes

Highly efficient enzyme membrane reactors can be also produced by immobilizing enzymes in membranes or in hollow fibers. For example, enzymes can be confined in the porous support matrix of an asymmetric capillary membrane, while substrate-containing solution flows through the fiber lumen. The dense skin layer at the lumen wall should be impermeable to the enzyme molecules. The latter diffuse through the inner wall of the fiber to the enzyme into the spongy part, where the conversion takes place. Applied transmembrane pressure and axial flow rate are parameters that contribute to control of the reactor performance. The development of improved immobilization techniques permits design of continuous flow reactors where biocatalysis could be achieved without enzyme loss in the effluent stream. Enzymes are also more protected and less exposed to denaturation.
In most of the traditional immobilization procedures,however, the contributions of recent progress in membrane technology have been very limited. The preparation of enzyme membranes on a large scale for industrial processes, in which selective mass transfer across the artificial membranes is combined with specific chemical reactions,would require low membrane cost and standard preparation procedures. Two immobilization procedures have been recently studied in our laboratory, which might accomplish those requirements. Gelled enzyme membranes,involving labile immobilization at the membrane-solution interface, can result

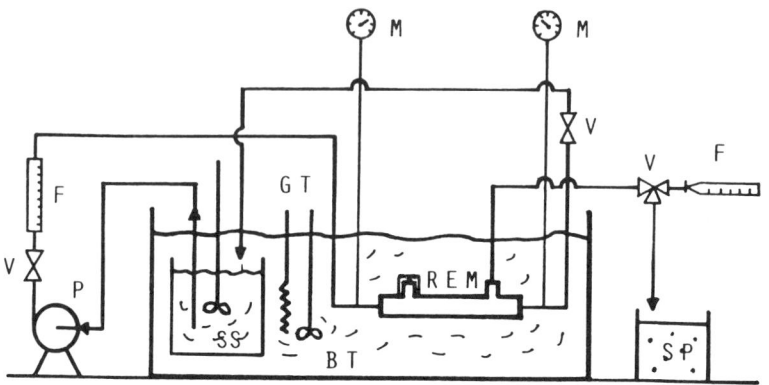

FLOW SHEET OF LABORATORY EXPERIMENTAL PLANT

Figure 3. Flow sheet of laboratory experimental plant. REM, capillary enzyme membrane reactor; SS, substrate reservoir, SP, permeate reservoir.

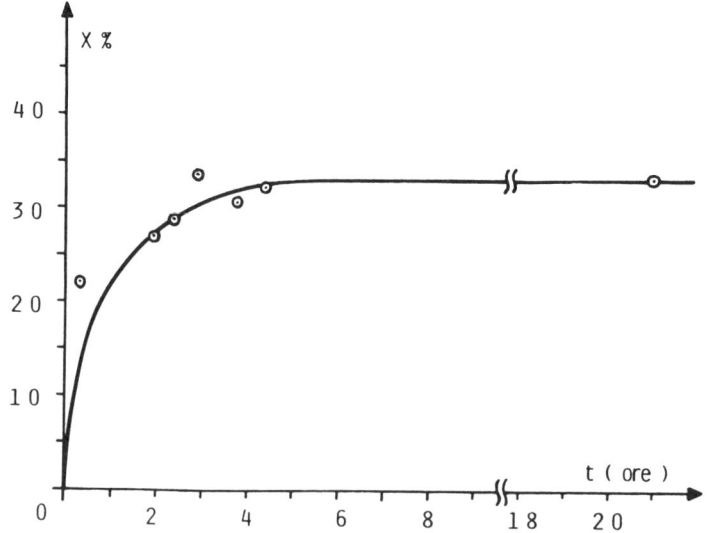

Figure 4. Triglycerides degree of conversion with time. pH = 6, olive oil 3%, lipase 30 mg in 500 ml; axial flow rate 960 ml/min.

from concentration polarization phenomena. Both in batch unstirred ultrafiltration processes and in UF processes with continuous recirculation of the substrate solution along the membrane, an appropriate amount of enzyme can be totally or partially immobilized in "gel" form on the pressurized face of the membrane, depending on the detailed fluid dynamics (_7_). Such a dynamically formed gelled enzyme membrane formation technique for acid phosphatase, urease, ß-galactosidase, malic enzymes, lipase, etc., has been studied. From the experimental results it appears that enzymes forming a gel layer on a UF membrane retain their activity and their stability is increased. The method has been carried out using flat membranes and tubular membranes in a continuous recirculating system. The polymeric material forming the supporting membrane does not significantly influence the enzyme stability. The ultrafiltration membrane cut-off and the membrane morphology, on the contrary, appear to be the controlling factors. The technique has been shown to be useful also for studying the kinetic behaviour of immobilized allosteric enzymes (_8_). The more traditional techniques in fact, lead to a decrease of the specific activity or to a deactivation of these enzymes, which are particularly sensitive to conformational transitions induced by specific ligands or environmental constraints. With this technique the enzyme should be immobilized without significant changes in the enzyme micro-environment; moreover, the situation is particularly favorable because of the high enzyme concentration in the gel. The possibility of using traditional ultrafiltration and reverse osmosis membranes filled with biocatalyst might represent a significant improvement in the development of membrane technology and enzyme engineering. However, the preparation at industrial scale of UF and RO membranes filled with enzymes or whole cells, using traditional techniques, has been limited by the need for non-aqueous solvents, in the casting solutions and high temperature annealing, which destroys the catalytic properties. The recent isolation of "Solfolobus Solfataricus" (_9_), an extreme thermophile growing optimally at 80 °C and whose enzymes are generally stable to protein denaturating agents, offers an interesting opportunity for using the phase inversion technique for the preparation of UF and RO membranes filled with whole cells as the enzyme source. S.Solfataricus contains enzymes of industrial interest, such as ß-galactosidase and malic enzymes, for example. Artificial membranes filled with S.Solfataricus have been prepared by several methods. Cellulose acetate and poly-

sulfone were used to obtain asymmetric membranes by the phase inversion technique (10). Albumin and glutaraldehyde were used for cell immobilization in membranes by the co-crosslinking method.
A hydrophilic polyisocyanate was used to prepare porous polyurethane structured foams in thin films. The use of liquid polyurethane prepolymer, which contains at least two free isocyanate groups per polymer molecule, has been recently suggested as an immobilyzing agent. Hydrophilic porous films and tubes can be prepared with this material, containing immobilized biologically active compounds (11). Physical entrapment and specific reaction between free isocyanate groups and protein amino-groups contribute to the "immobilization".
The physico-chemical properties of trapped-cell ß-galactosidase activity in the above models were similar to those shown by the enzyme in the free cell. At the optimal pH the ß-galactosidase exhibited maximal activity at about 100 °C and appeared stable for up to 24 hr at room temperature and a pH of 3-8. Incubation of trapped-cell ß-galactosidase for up to 24 hr at room temperature with organic solvents did not cause any loss of activity. After 8-9 months of wet storage at 4 °C, no decrease of enzymatic activity was observed. Cell entrapment imparted a significant increase in enzymatic activity in comparison with intact free cells. This effect may be a consequence of cytoplasmic membrane permeabilization of the microorganism caused by by the entrapment procedures. The increase in enzymatic activity was 35-fold greater for the polyurethane system than in the other membrane preparations. The permeate flow rate, ß-galactosidase degree of conversion, and stability of the system were studied in the range of 70-85 °C. Those membranes have been also prepared in capillary configuration from a dope consisting of a mixtures of polysulphone, polyvinylpyrrolidon and N-N dimethylacetamide. Before spinning the dope lyophilized bacteria, 3% w/v, were added to the mixture (Figure 5).
Membranes were formed according to the phase inversion technique, (12) spinning the dope in the equipment in Figure 6 : water was used as non-solvent agent to promote membrane formation.
Microphotographs of the membranes, Figure 7, give clear indications on the distribution of bacteria both in the wall of the membranes and underneath the dense layer. The capillaries still exhibit the dense internal skin layer typical of asymmetric membranes, and a supporting finger structure (1 = 10-20 μm) where bacteria are preferentially allocated.

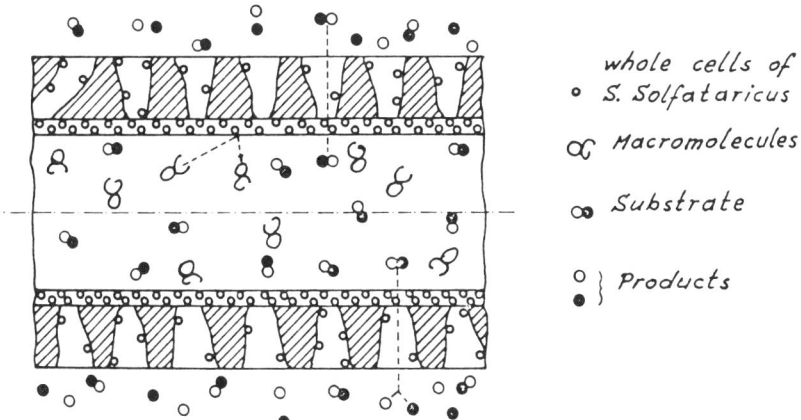

Figure 5. Asymmetric enzyme capillary membranes prepared by phase invertion method.

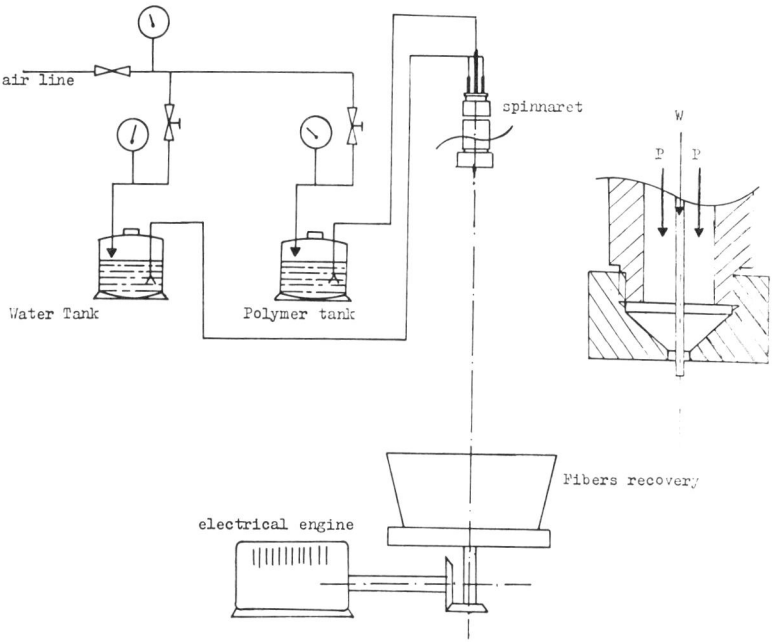

Figure 6. Flow sheet of the spinning system for capillary membrane formation.

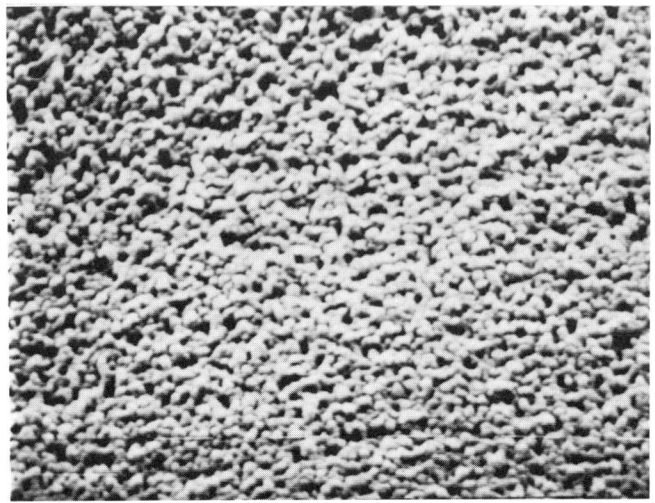

Figure 7a. SEM picture of the dense skin of an asymmetric capillary membrane with entrapped cells. Reproduced with permission from Ref. 14.

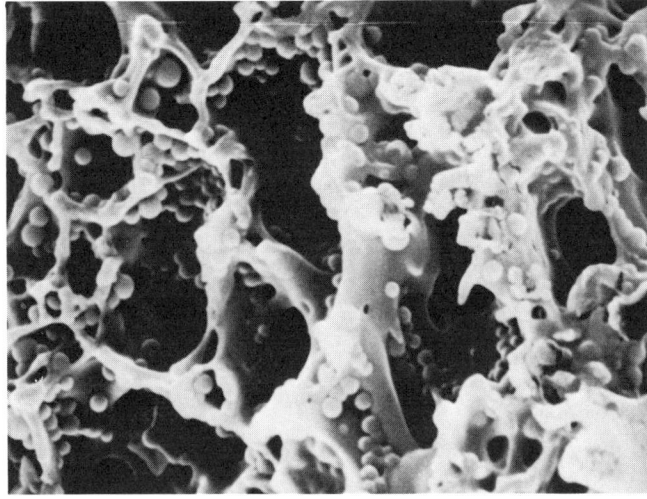

Figure 7b. SEM picture of the porous sublayer of an asymmetric capillary membrane with entrapped cells. Reproduced with permission from Ref. 14.

Mechanical properties of the membranes were preliminarly tested and compared to those exhibited by cell-free polysulphone fibres. The Young modulus, E wet, and the ultimate properties of the membranes were estimated by a stress-strain analysis carried out on a Instrom Universal Tester. The average value of the Young modulus was found lower by a factor of about 2.5 relative to the average value of the cell-free fibres.

The catalytic activity of membrane entrapped cells was assessed evaluating the rate of glucose production, defined as permeate flow rate times glucose permeate concentration at different lactose concentration, and transmembrane pressures.

Referring to product concentration in the permeate stream, cytoplasmatic ß-galactosidase in cells exhibits an apparent Michaelis-Menten kinetic behaviour, Figure 8. The apparent Michaelis constant increases from 3.7 to 19.2 mM as the pressure increases. Maximum glucose production at 70 °C ranges from 20.4 to 34 moles/hr, at 0.04 and 0.055 atm, respectively.

In terms of stability the ß-galactosidase of the microorganisms shows to be stable up to one year without any appreciable loss of activity, when stored, and up to three months, at least, during continuous operation.

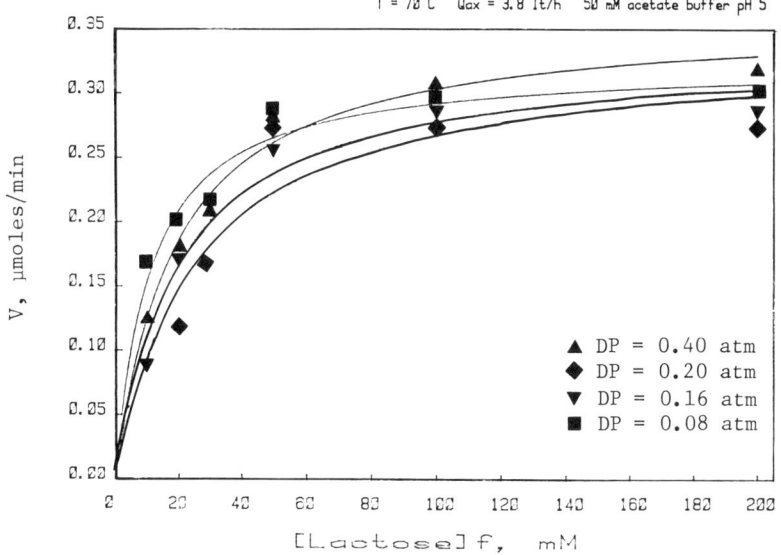

Figure 8. Glucose production rate vs. lactose feed concentration. T = 70 °C. Reproduced with permission from Ref. 13.

From the point of view of mechanical properties, performances of capillary membranes charged with cells are almost comparable to those of bacteria-free ones. The interesting conversions observed in lactose hydrolysis and the remarkable stability of immobilized bacterial ß-galactosidase encourage further studies for the development of an enzyme membrane reactor oriented to possible industrial applications.

Literature Cited

1. "Impact of Applied Genetics",OTA NR-132 April 1981.
2. Michaels,A.S. Desalination 1980,35,329.
3. Michaels,A.S.; Matson,S.L. Desalination 1985,53,231-258.
4. Drioli,E.; Serafin,G.; Rigoli,A. presented at First Engineering Foundation Conference " Advances in Fermentation Recovery Process Tech." Banff June 6-12 (1981) unpublished results.
5. Leuchtenberger,W.;Karrenbauer,M.;Plcker,U. "Scale up of an Enzyme Membrane Reactor Process for the Manufacture of L-enantiomeric Compounds" Report from Degussa AG, D-6450 Hanau I, FDR.
6. Molinari,R.;Drioli,E. Proc.Nat.Congr.Ind.Chem.Div.Sci., Siena 10-12 June 1985
7. Drioli,E.;Scardi,V. J.Mem.Sci. 1976,1,237-248.
8. Rossi,M.;Nucci,R.;Raia,C.A.;Molinari,R.;Drioli,E. J.of Mol.Cat. 1978,4,233.
9. De Rosa,M.;Gambacorta,A.;Esposito,E.;Drioli,E.;Gaeta,S. Biochimie 1980,62, 517
10. Drioli,E.;Iorio,G.;De Rosa,M.; Gambacorta,A.;Nicolaus,B. J.Mem.Sci. 1982,11,365-370
11. Drioli,E.;Iorio,G.;Santoro,R.;De Rosa,M.;Gambacorta,A.;Nicolaus,R. J. Mol.Cat.1982,14,247.
12. Catapano,G.;Iorio,G.;Drioli,E.;Filosa,M. "Capillary Membrane Bioreactors with Entrapped Whole Cells : a Theoretical Model" submitted for publication.
13. Drioli,E.;Iorio,G.; Catapano,G.; De Rosa,M.; Gambacorta,A. J.of Mem.Sci. in press.
14. Drioli, E.; et al. La Chimica E L'Industria 1985, (67)11, 617-622.

RECEIVED March 26, 1986

Liquid Emulsion Membranes and Their Applications in Biochemical Separations

M. P. Thien, T. A. Hatton, and D. I. C. Wang

Department of Chemical Engineering, Massachusetts Institute of Technology, Cambridge, MA 02139

> Applications of liquid emulsion membranes (LEMs) to biomedical and biochemical systems are reviewed and other potential applications identified. The LEM-mediated downstream processing of small, zwitterionic biochemicals (e.g. amino acids) is examined using chloride ion counter-transport to separate and concentrate the amino acid phenylalanine from stimulated fermentation broth. The effect of agitation rate and osmotic swelling of membranes on separation is shown to be significant.

Fermentation technology has progressed much in the past decade. Advances in genetic engineering, coupled with better analytical techniques and more efficient fermentation instrumentation, have made possible the large-scale production of many complex and exotic biochemicals. Unfortunately, advances in separation and purification technology have not kept abreast of those made in fermentation. Most downstream processing requires a number of traditional and inherently batch unit operations, operations which are generally costly and inefficient. One new technique, the use of liquid emulsion membranes, presents great potential as an efficient means of separating and concentrating fermentation products. In this paper, a review of LEM technology is presented with some emphasis given to the limited work done in the area of biotechnology. As a specific example of the potential for LEMs in biotechnology, we summarize our work on the recovery of the amino acid phenylalanine from simulated fermentation broth.

Concept

Since they were first developed in 1967 (1), liquid emulsion membranes (LEMs) have been used in a variety of separations (see reviews by Frankenfeld and Li (2), Marr and Kopp (3), and Way et al. (4)). Conceptually, LEMs are quite simple. They consist of an emulsion of two immiscible phases which, once formed, is subsequently dispersed in a third (continuous or "exterior") phase (see Figure 1). As is seen in Figure 1, one of the two phases which comprise the emulsion is completely encapsulated by the other phase. This encapsulated "interior" or "dispersed" phase thus never directly contacts the continuous or external phase. In most applications the interior and exterior phases are aqueous solutions. The non-dispersed emulsion phase (usually an oil, making the resulting emulsion a "water-in-oil," W/O, emulsion) acts as a membrane between the interior and exterior phases and is thus called the membrane phase.

Separations using LEMs can be effected in two different ways (5). The first of these (called TYPE I transport) is a simple diffusion process in which the solute partitions into the membrane phase from the exterior phase, diffuses across the membrane to the dispersed interior phase droplets, and partitions into the interior phase. A reaction takes place in the internal phase which converts the solute into a species which is incapable of partitioning back into the membrane phase. TYPE I transport is limited to uncharged solutes only, since only uncharged solutes will be able to favorably partition into the membrane phase.

The second type of transport process, TYPE II or facilitated transport, is shown in Figure 2 using L-phenylalanine, an amino acid, as the solute to be separated and concentrated. The solute, due to its charge (the phenylalanine is separated while in its anionic form), cannot partition into the oil phase by itself. Consequently, a non-water soluble "carrier" molecule, usually an ionic surfactant consisting of a hydrophobic section and an univalently charged functional group, is added to the membrane phase. The example in Figure 2 uses a quaternary ammonium salt as the carrier. In order to remain in the oil/membrane phase, the carrier coordinates with a counter-ion (chloride, in this case). If the carrier is already complexed with the interior phase counter-ion before separation begins, the carrier-ion complex diffuses across the membrane to the exterior/membrane phase interface. At this interface an ion-exchange reaction takes place in which the solute is exchanged for the counter-ion. This reaction is driven either by "mass action" (i.e. there is very little counter-ion in the exterior phase compared to the solute), or by the carrier's high affinity for the solute. The carrier-solute complex then diffuses back across the membrane to the interior phase where the solute is exchanged for the counter-ion (usually present in great excess). If the counter-ion in the membrane is concentrated enough, the transport of solute will be driven by the counter-ion gradient (in a process akin to active transport across cell membranes). Since the volume of the interior phase can be made much smaller than that of the exterior phase, separation and concentration occur simultaneously as the solute is transported from exterior to interior phase.

While providing a means of separating and concentrating a given solute, LEMs are not without potential process difficulties. One such difficulty is that of mechanical stability. While considered as a significant concern by some (2,7,8), proper formulation of the membrane phase can almost eradicate any process difficulties associated with membrane breakage. Another, and perhaps much more

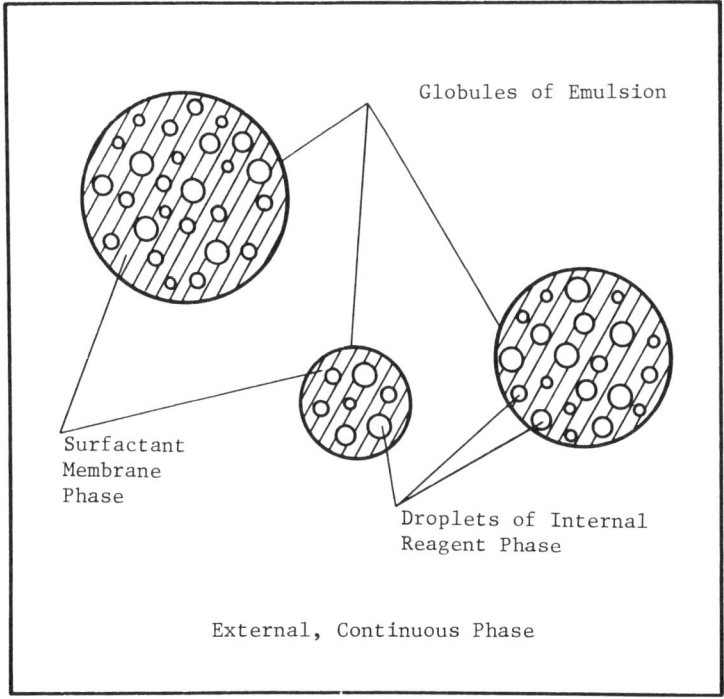

Figure 1. Schematic diagram of a liquid emulsion membrane (LEM) system.

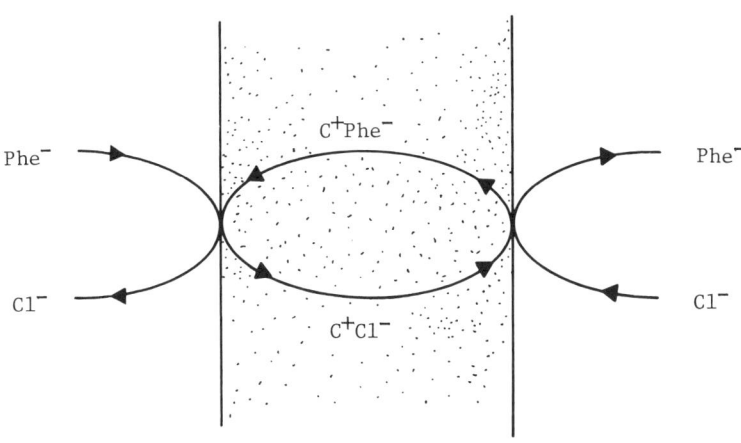

$C^+ = (C_8 - C_{10})_3 - N^+ - CH_3$

Figure 2. Mechanism for type II (facilitated) transport in a phenylalanine/chloride system.

of a significant concern, is that of membrane swell. Osmotic swell is a process by which water is transferred into the interior aqueous phase via the diffusion of hydrated surfactant molecules. Much like a regular TYPE II transport process, it is believed (6) that water is transported across the membrane due to the osmotic gradient established by the concentration of internal phase reagent. Although seldom mentioned in the literature (7-9), swelling has been reported by some to be as high as 150% of the interior phase volume (10). Colinart et al. (6) undertook a mechanistic study showing that, since practically all surfactants hydrate to some extent, swelling can be anticipated for all LEM formulations and could be a significant problem in LEM separations.

Applications

In spite of the potential problems of swelling and breakage, LEMs have been tested on the pilot plant scale with good results (2,9,11, 12), and will soon be used for the commercial separation of zinc from Viscose wastewater from rayon and cellophane processing (13). It should be noted that these pilot plant studies indicated that LEM processes were as economically advantageous, if not more so, than currently employed solvent extraction and conventional ion exchange techniques. Unfortunately, LEM-mediated separations of biochemicals have not been carried out on a pilot plant scale.

LEMs have been applied for the separation and recovery of a host of different compounds. Previous efforts have been primarily focused on the recovery of metal ions from aqueous solutions (including copper, zinc, chromium, mercury, uranium, nickel, and iron; (3)) and the removal of organic compounds from wastewater (14-17).

In terms of the amount of literature developed, biochemical separations have been largely ignored by those in the field of LEM-mediated separations. One application that has enjoyed some experimental scrutiny is that of the use of LEMs in drug delivery and overdose prevention systems. They have been used to separate or release several different types of drugs including acetylsalicyclic acid (18), phenobarbital (19), and several barbiturates (20,21).

LEM systems have also been shown to be successful in separating commodity-type biochemicals such as propionic acid (10) and acetic acid (10,22) and have been used for the preparation of L-amino acids from racemic D,L mixtures by means of enzymatic hydrolysis of amino acid esters (23). In addition to biochemical separations, the work of Mohan and Li showed that enzymes could be encapsulated in liquid emulsion membranes with no deleterious effect on enzyme action (24). Later work by these authors indicated that encapsulated live cells could remain viable and function in the LEM interior phase for period as long as five days (25).

These various uses of liquid emulsion membranes show the versatility of LEM-mediated separations and point to possible applications of liquid emulsion membranes in the biochemicals field. These applications include the *in-situ* recovery of fermentation products. This would be particularly useful in fermentations where the major product inhibits growth or production rates. Examples include those systems which produce organic acids such as acetate and lactate. Perhaps the most useful application would be the downstream processing of biochemicals that can only be economically separated by costly ion exchange techniques. One example of this last category of applications is that of the downstream processing of amino acids. This application has been thoroughly investigated by the authors and our studies are summarized below. A more extensive analysis of our results will be given in a later publication (26).

Amino Acid Recovery Using Liquid Emulsion Membranes

Amino acids find a variety of commercial applications, including use as livestock feed supplements, nutritional supplements, and raw materials for speciality chemicals (e.g. glutamic acid for monosodium glutamate, MSG). Most of these amino acids are produced in commercial quantities by microbial fermentation. Unfortunately, some amino acids must be separated by complicated and costly ion-exchange schemes due to low fermentation titers and very low solubility in organic media (solubilities low enough to preclude the use of solvent extraction as a means of separation). Subsequent separation and purification costs may comprise as much as 80% of total production costs for these amino acids. One such product, phenylalanine, has recently enjoyed great commercial success as one of the raw materials of the non-nutritive sweetner aspartame (Nutrasweet). Phenylalanine having somewhat lower fermentation titers than typical biochemical commodities, coupled with the current commercial interests in phenylalanine and its ability to be easily measured by UV spectrophotometry, make phenylalanine a good model compound for testing new biochemical separation techniques. This being the case, we have chosen to test the applicability of LEM-mediated separations to the problem of the downstream processing of phenylalanine from fermentation broth.

Experimental System

The system examined in this study was that of a simple phenylalanine/chloride counter-transport system in which the interior phase was concentrated in chloride anion and the exterior phase contained concentrations of phenylalanine in the range of commercial fermentation titers. Since phenylalanine (like all α-amino acids) is predominantly a zwitterion at pH's between 3.5 and 9.0 (and is thus unable to be transported through the membrane), the pH of the exterior phase was kept above pH 10.0 to insure that the amino acid was present in its transportable anionic form. As mentioned above, the mechanism for this system is shown in Figure 2.

Experiments typically consisted of dispersing 70 milliliters of a 2 molar solution of sodium chloride (NaCl) in 100 milliliters of an oil phase. This oil phase consisted of a paraffinic solvent (Solvent 100 Neutral, Exxon Chemical Company), a nonionic emulsion-stabilizing surfactant (Paranox 100, Exxon Chemical Company), a cationic "carrier" - a tri-capryl quaternary ammonium salt (Aliquat 336, Henkel Corporation), and a co-surfactant for the carrier molecule (decyl alcohol, Sigma Chemical). The aqueous phase was slowly added to the oil phase under intense shear provided by a stator-type homogenizer. All emulsions had similar internal droplet size and size distributions as analyzed by a Horiba CAPA 500 particle size distribution analyzer. The exterior phase consisted of 700 milliliters of 11 g/l L-phenylalanine (Sigma Chemical) brought to pH 11.0 by potassium hydroxide addition. The emulsion formulation and phase volumes were not optimized for separation performance.

Batch separation experiments were carried out in a baffled glass 2 liter vessel at 25°C. A measured amount of emulsion was poured into the reaction vessel, a measured quantity of exterior phase was added to the vessel and the agitation and timer were turned on. Vessel contents were agitated using six marine-type impellers. Impeller RPM was monitored by strobe light. Samples were taken throughout the experiment via a sampling port at the bottom of the vessel. Emulsion was quickly separated from the exterior phase of samples and discarded.

The exterior phase was analyzed for phenylalanine concentration and pH. All sample volumes were recorded and used for mass balance determination. Phenylalanine was measured spectrophotometrically at λ_{max} = 257.5 nm. Changes in interior phase volume were calculated using material balances. All material balances closed to within 2%. Interior phase concentrations were estimated by the use of material balances and exterior phase concentrations. The interior phase components of several representative emulsions were measured by analyzing the interior phase components after thermally demulsifying the emulsion samples. These measurements agreed with estimates to within 10%.

Results

Separation results for a typical membrane formulation used in our studies are shown in Figure 3. The results, presented in a concentration profile format, indicate that initial separation is fast (when the driving force, the chloride gradient, is greatest) but slows down as transport continues and chloride is transported to the exterior phase. It should be noted that, although the profile gives a rough idea of separation performance, it does not indicate the respective volumes of the various phases and thus does not account for the effects of membrane swell or breakage. The incidence and ramifications of these effects will be discussed later.

The agitation delivered to the reaction vessel is an important parameter in LEM separations. The agitation rate not only is the dominant factor in determining the significance of mass transfer resistance external to the emulsion globules, but it also determines the emulsion globule size, and thus the area available for transport. The effects of agitation on LEM amino acid separations are shown in Figures 4 and 5. The traditional profiles in Figure 4 indicate that the mass transfer rate is increased with increases in agitation.

Figure 5 shows changes of interior phase volume ("% swell") based on initial interior phase volume) and the estimated internal phase phenylalanine concentration as a function of agitation speed. This figure indicates that, contrary to expectations based on the results given in Figure 4, the best separation (in terms of concentrating phenylalanine) occurs at 400 RPM where changes in membrane volume are at a minimum.

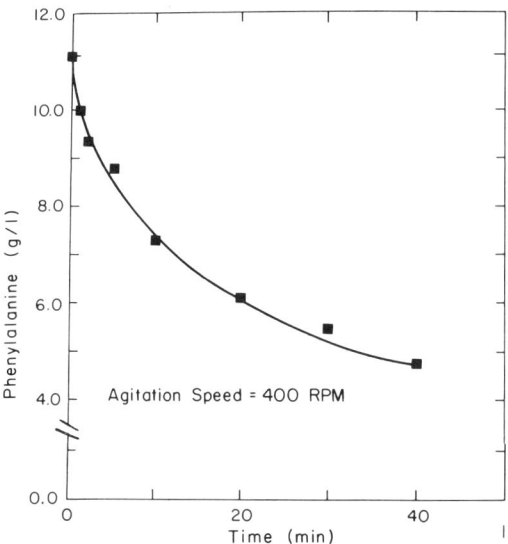

Figure 3. Chloride countertransport system: typical LEM separation.

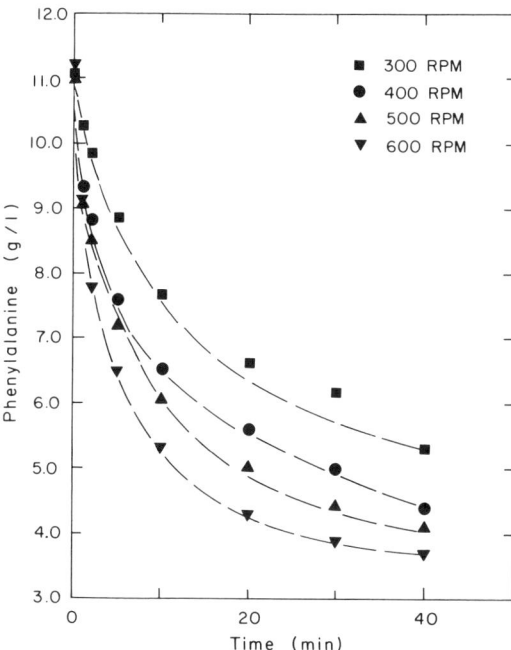

Figure 4. Effect of agitation speed on LEM separation.

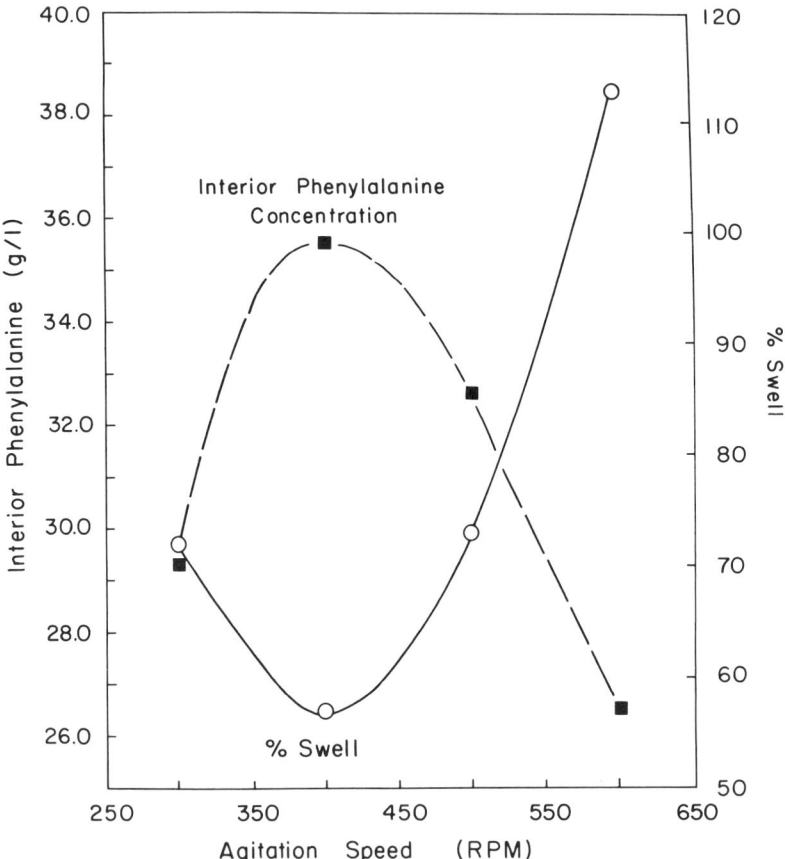

Figure 5. Effect of agitation speed on swell and interior phenylalanine concentration.

Discussion

The above work shows that amino acids can indeed be separated and concentrated using liquid emulsion membranes. The results also indicate some very important concerns when considering the use of LEM-based separation processes. Contrary to traditional methods of evaluating LEM performance, the parameter of direct interest, that is the interior phase concentration, must be examined. The effects of swelling and breakage, due to their potential significance, must always be considered seriously. Several different strategies could be formulated to minimize the deleterious effects of swelling in the above counter-ion process. Further improvements, such as an amino acid-specific reaction in the interior phase (to maintain a maximum driving force) could be used to significantly improve the performance of LEM systems for amino acid separations.

A schematic of one possible commercial LEM recovery scheme for amino acids which takes advantage of their unique solubility properties is shown in Figure 6. Once the amino acid has been concentrated in a basic interior phase, the emulsion is broken and the interior phase neutralized with the acid of the counter-ion to approximately pH 7.0. It has been found that amino acids have a much higher solubility in aqueous solutions of extreme pH than in the broad isoelectric range centered around pH 7.0 (27). Neutralizing the pH of the interior phase should cause the spontaneous crystallization of the amino acid, leaving only that amount of amino acid that is soluble in the isoelectric range in solution. The resulting neutralized solution may then be used again as the interior phase after salt and base is added.

Conclusions

The versatility of LEMs is clear. From the encapsulation of living cells to the removal of toxic or inhibiting substances, and in their use as a downstream process, liquid emulsion membranes remain a powerful and, as of yet, virtually untapped resource for biochemical engineers. The ability of LEMs to separate and concentrate amino acids demonstrated here gives strength to this observation, and it is anticipated that these systems will enjoy increasing attention in the years to come.

Acknowledgments

This work was funded in part by a Monsanto Junior Faculty Award to T.A. Hatton. We express our gratitude to Karen Lee, Linda Marinilli, and Todd Renshaw for their valuable technical assistance.

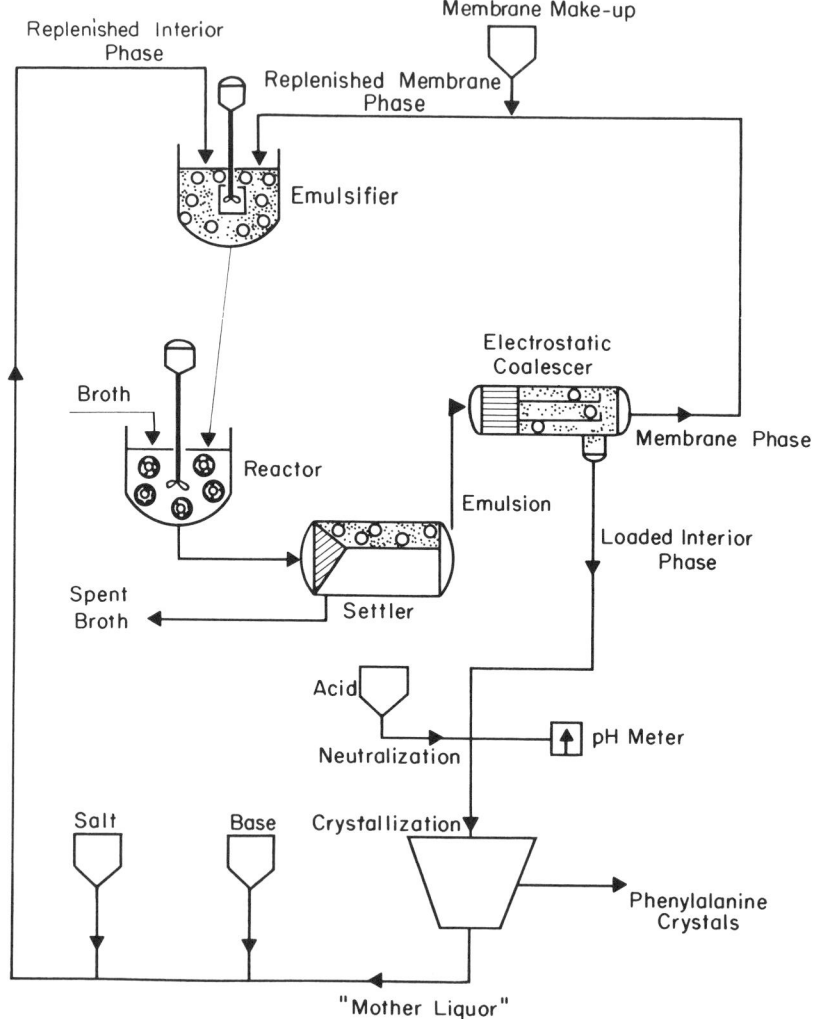

Figure 6. Process diagram for LEM-based recovery of amino acids.

Literature Cited

1. Li, N. N. U.S. Patent 3 410 794, 1968.
2. Frankenfeld, J. W.; Li, N. N. Recent Dev. in Sep. Sci. 1978, 5, 285.
3. Marr, R.; Kopp, A. Int. Chem. Eng. 1982, 1, 44.
4. Way, J. D.; Noble, R. D.; Flynn, T. M.; Sloan, E. D. J. Mem. Sci. 1982, 239.
5. Matulevicius, E. S.; Li, N. N. Sep. Pur. Meth. 1975, 4, 1, 73.
6. Colinart, P.; Delepire, S.; Trouve, G.; Renon, H. J. Mem. Sci. 1984, 20, 167.
7. Volkel, W.; Halwachs, W.; Schugerl, K. J. Mem. Sci. 1980, 6, 19.
8. Martin, T. P.; Davies, G. A. Hydrometallurgy 1977, 2, 315.
9. Cahn, R. P.; Frankenfeld, J. W.; Li, N. N.; Naden, D.; Subramanian, K. N. Rec. Dev. in Sep. Tech. 1982, 6, 51.
10. Terry, R. E.; Li, N. N.; Ho, W. S. J. Mem. Sci. 1982, 10, 305.
11. Li, N. N.; Shrier, A. L. Rec. Dev. in Sep. Tech. 1972, 1, 163.
12. Hayworth, H. C.; Ho, W. S.; Burns, W. A. Sep. Sci. Tech. 1983, 18, 6, 493.
13. Protsch, M.; Marr, R. Proc. Int. Sol. Ex. Conf., 1984, p. 66.
14. Kitigawa, T.; Nishikawa, Y.; Frankenfeld, J. W.; Li, N. N. AIChE J. 1982, 28, 4, 662.
15. Teramoto, M.; Takihana, H.; Shibutani, M.; Yuasa, T.; Miyako, Y.; Teranashi, H. J. Chem. Eng. of Japan 1981, 14, 122.
16. Park, H. S.; Yoo, J. H.; Suh, I. S.; Han, P. S.; Kay, W. K.; Burgaud, M.; Fleury, M. J. Proc. Int. Sol. Ex. Conf., 1984, p. 288.
17. Baird, R.S.; Bunge, A. L.; Noble, R. D. "Batch Extraction of Amines Using Emulsion Liquid Membranes-Importance of Reaction Reversibility," to be published.
18. Yang, T. T.; Rhodes, C. T. J. Appl Biochem. 1980, 2, 7.
19. Chilamarki, R. N.; Rhodes, C. T. J. Appl Biochem. 1976, 2, 405.
20. Chang, C. W.; Fuce, G. C.; Frankenfeld, J. W.; Rhodes, C. T. J. Pharm. Sci. 1978, 67, 1, 63.
21. Parkinson, G.; Short, H.; McQueen, S. Chem. Eng. 1983, 8 22, 22-7.
22. Larson, K. M.; Hanna, G.; Hanson, S.; Way, J. D. AIChE Symp. San Francisco, CA, 1984.
23. Scheper, T.; Halwachs, W.; Schugerl, K. Int. Sol. Ex. Conf., ISEC, 1983.
24. Mohan, R. R.; Li, N. N. Biotech. and Bioeng. 1974, 16, 513.
25. Mohan, R. F.; Li, N. N. Biotech. and Bioeng. 1975, 17, 1137.
26. Thien, M. P.; Hatton, T. A.; Wang, D. I. C. 1986, in preparation.
27. Needham, T. E.; Paruta, A. N.; Gerraughty, R. J. J. Pharm. Sci. 1974, 60, 2, 258.

RECEIVED March 26, 1986

7

Use of Aqueous Two-Phase Systems for Recovery and Purification in Biotechnology

Bo Mattiasson and Rajni Kaul

Department of Biotechnology, Chemical Center, University of Lund, S-221 00 Lund, Sweden

> Aqueous two-phase systems are generated by mixing aqueous solutions of two water-soluble polymers, or a polymer and a salt. These systems offer extremely mild conditions for separation of cells, organelles, proteins and other biomolecules, in biochemical processes. Considerable attention has been directed towards the use of the two-phase systems in several areas of biotechnology. The present paper summarizes the state of the art concerning extractive bioconversions for production of small as well as macromolecules, and protein purification using aqueous two-phase system.

The well documented phenomenon of separation of an aqueous solution of two different water-soluble polymers into individual phases, during recent years, has shown widespread potential in biotechnology (1). A number of polymers have been employed for the preparation of these bi-phasic systems. (2). The most commonly used systems have been those of poly(ethylene glycol) (PEG) and dextran. The molecular weight of the polymers used plays an important role in determining the characteristics of the phase system. Phase systems formed by mixing one polymer and a high concentration of certain salts in aqueous solutions, have also been reported. (2).

Aqueous two-phase systems are characterized by a high water content, which makes them compatible with the biological material. The distribution of biomolecules between the phases is determined mainly by their surface properties and the composition of the phase system, and is denoted by partition coefficient, K_{part}, defined as the ratio of its concentration in the top and bottom phase, respectively. The partitioning is independent of the absolute concentration of the substance over a fairly wide range. Any substance prefers a phase where a maximum number of interactions are possible; these could be related to electrical, hydrophilic, hydrophobic and conformational forces. Reports in the literature have shown that different characteristics of the system can be manipulated in order to achieve the desired partitioning effect (2). Though, as a general rule, small uncharged molecules are distributed quite evenly between the phases (i.e. $K_{part} \sim 1.0$). Particles, such as cell organelles or cells, normally partition between the interface and one of the bulk phases. The distribution of macromolecules like proteins, is quite variable, and also shows great sensitivity to changes in the composition of the phase system.

The use of the aqueous two-phase systems in biotechnology basically exploits this varying distribution of biomaterials between the phases. These systems can be buffered and are suitable for carrying out bioconversions. The phase polymers have also been shown to have a stabilizing influence on biocatalysts (3,4); the latter are, in a way, temporarily immobilized within liquid droplets. The different areas in which the two-phase systems have shown potential include extractive fermentations, purification of biomolecules, cells, membranes and organelles, and biological binding assays (1). However, most of the systems reported so far are based on highly purified phase components, the costs of some of which poses a severe limitation on the scale up of the processes. Nevertheless, there has been a serious attempt in employing cheap phase components, for example, in purification of proteins on large scale (5,6).

This paper is meant to be an overview of the developments in the use of aqueous two-phase systems in the recovery of products of fermentation, and also purification of proteins and other biomolecules. It also touches upon the means to improve the economics of the process to make it industrially feasible. It has been realized that the closer integration of bioconversion processes with downstream technology is essential for quicker and more economical enrichment of the product. Initial efforts have already been taken in this direction, some of which are discussed here.

Extractive bioconversions in aqueous two-phase systems

Product inhibition seems to be a general phenomenon in traditional fermentation processes, which are, therefore, characterized by dilute product streams. The need to obtain high product concentrations in the fermentation broth is essential both for its yield and recovery. The major improvement towards achieving this goal has been focused, in recent years, on genetic manipulation of microorganisms, as a result of which strains have been obtained that are tolerant to high concentrations of the product. At the same time, however, the intensification of the product recovery processes is also required, which has until now been carried out separately in several stages, often giving low yields. The continuous removal of the product by coupling the extraction step with that of fermentation (extractive bioconversion) would help in minimizing the inhibitory effects and/or degradation phenomena. This may also lead to the partial enrichment of the product, which will facilitate the eventual purification steps as well.

In an aqueous two-phase system, if the biocatalytic reaction takes place in one of the phases, while the products are either evenly distributed, or preferentially partitioned to the other phase, an ideal situation for extractive conversion could be visualized (Figure 1). Aqueous two-phase systems are characterized by the requirement of minimal energy to create a finely dispersed emulsion resulting in a large interface and high mass transfer, which is especially advantageous in reducing microenvironmental product inhibition (1). Recently, studies have been performed on the extractive fermentation of chemicals, of interest to the biological and chemical industry, in two-phase systems.

Production of bulk chemicals. The production of solvents is normally characterized by a general inhibition phenomenon which has been mainly attributed to the changes in membrane permeability, or to the toxic effects on the metabolic pathway. Aqueous two-phase systems have been shown to be effective as media for the extractive fermentation of a number of solvents which include ethanol, acetone-butanol and acetic acid (3). Improved productivity has been achieved in most of the cases as compared to the conventional fermentations, which is significantly due to the elimination of product inhibition. However, there is an indication that changes in the microenvironment of the microbial cells due to the presence of non-metabolizable polymers could also contribute, in the initial phases, to the increased production. The addition of PEG and dextran to a growth medium, for instance, was shown to give increased initial ethanol yields, as a result of decrease in the chemical potential of water (8).

Much effort has gone, in recent years, in setting up alcoholic fermentations based on immobilized cell technology (9). Some of the systems have proved to be highly productive, but are faced with drawbacks of leakage of cells, and sterical hindrances. Fermentation in two-phase system, on the other hand, has been successfully carried out with macromolecular substrates such as starch and cellulose (7,10). It is also easier to control a reaction system involving a number of enzymes, in a two-phase system as compared to the immobilized systems; for example, there is a possibility to add more of the labile catalyst during the continuous operations.

All the extractive fermentation processes studied in the two-phase system have employed PEG and dextran as the phase polymers (Table I).

The microbial cells employed for the conversion have been seen to be enriched in the dextran rich bottom phase and also, the interface. The macromolecular substrates are also found located in the bottom phase. In fact, in one of the earlier studies on ethanol fermentation, starch alone constituted the lower phase of a two-phase system (3). The solvent molecules are rather evenly distributed between the two phases. However, the partitioning behaviour of the solvent molecules can somewhat be changed by variations in the phase composition. Furthermore, a significant extraction of the solvent into the upper phase could be achieved by increasing the top to bottom phase volume ratios (1,11). Semicontinuous batch fermentations have been performed with the cells being recirculated in the bottom phase. After the conversion, the top phase with the product is removed and replaced by a fresh one along with more substrate. Alternatively, the phase is returned to the system after removal of the solvent e.g. by distillation (12,13). The

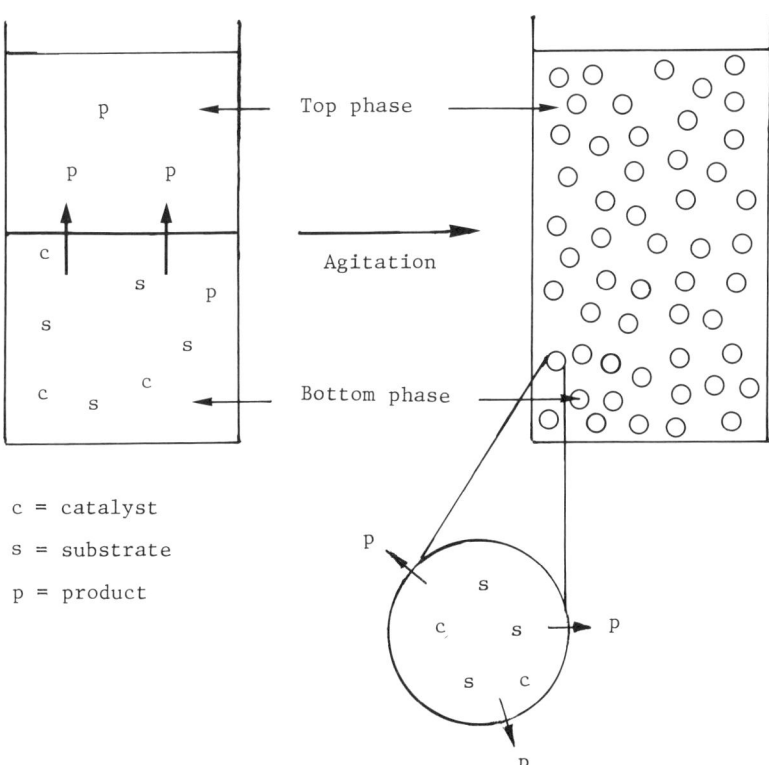

Figure 1. Principle for extractive bioconversion in aqueous two-phase systems, where the biocatalyst is temporarily immobilized in the droplets of one of the phases. Reproduced with permission from Ref. 7. Copyright 1984. Society of Chemical Industry.

integration of membrane technology with the phase system has been shown to be advantageous for the efficient recovery of the product during starch bioconversion (7). In this manner, any losses of the upper phase polymer and the biocatalyst are also reduced to a minimum. The repeated use and the increased operational stability of the biocatalyst helps in improving the economics of the process.

Table I. Bulk chemical production in aqueous two-phase systems

Bioconversion	Biocatalyst	Phase System	Reference
Glucose ⟶ ethanol	Saccharomyces cerevisiae	6% PEG 8000- 2% Dextran T500	(12)
Starch ⟶ ethanol	α-amylase; glucoamylase S.cerevisiae	5% PEG 20 M- 3% crude dextran	(7)
Cellulose ⟶ ethanol	cellulase, β-glucosidase + S.cerevisiae	PEG 8000- Dextran T40	(10)
Glucose ⟶ acetone-butanol	Clostridium acetobutylicum	25% PEG 8000- 6% Dextran T40	(13)
Glucose ⟶ acetic acid	Escherichia coli	6% PEG 8000- 7.5% Dextran	(3)

Reproduced with permission from Ref. 11. Copyright 1985, Verlag Chemie.

One of the common side effects observed during extractive bioconversion is the accumulation of unwanted by-products in the system which may affect the productivity during continuous operation (14). The build up of glycerol and other non-volatile products was shown to decrease the ethanol yields during repeated fermentations in a two-phase system (12). The problem was, however, solved by dialysing the fermentation broth and also adding more yeast cells. It appears that the combination of ultrafiltration with the phase system may circumvent the problem of by-product inhibition in most of the cases.

Even though aqueous two-phase sytems hold promise for bulk chemical production, their applicability on a large scale is not assured unless the phase components, at least, the fractionated dextran, are replaced by cheaper polymers, or technology is developed permitting full recycling of the polymers. These aspects are discussed later in the paper.

Production of fine chemicals. Inspite of the interest shown in the production of bulk chemicals in aqueous two-phase systems, the potential of these systems for fine chemical production has not yet been exploited. The only bioconversion reported has been the deacylation of benzyl penicillin to 6-amino penicillanic acid (15). Today, industrial deacylation is performed by penicillin acylase in an immobilized form. The productivity of the reaction in a two-phase system was found to be in the same range as that of immobilized system. But, the use of phase system may have an advantage over immobilized reactors, in that the inactivated enzyme may be replaced at any time during operation.

The slow development in the area of fine chemical production by enzymatic/microbial methods could be due to the fact that most of the compounds of commercial interest have poor solubilities in aqueous solutions. On this score, enzymatic synthesis has not been able to compete with traditional technology. However, recent studies on a few systems have shown that the presence of an organic solvent in a biocatalytic reaction medium offered some advantages with respect to solubility and downstream processing (16). The aqueous-organic solvent two-phase systems have provided effective reaction media for the extractive transformation of steroids (17). But this approach may not be applicable for processes involving living cells because of the denaturing effect of most of the solvents. This is in direct contrast to the two-phase systems, which are biocompatible and are characterized by a low surface tension between the phases.

The polymers generally employed in forming aqueous two-phase systems are quite hydrophilic as compared with organic solvents; there is, however, a marked difference in hydrophobicity of the phase polymers. Thus, by extraction into the more hydrophobic phase it may be possible to design bioconversion in aqueous two-phase systems of substances of low water solubility. To this aim, we examined the transformation of hydrocortisone to prednisolone by Arthrobacter simplex cells in 25% PEG 8000 - 6% Dextran T40 system (18). Our studies showed that while the cells preferred the bottom phase, the K_{part} of the steroid was in favour of the PEG rich upper phase ($K_{part} \sim 3.0$ and ~ 7.0 for hydrocortisone and prednisolone respectively). The reaction rate was found to be comparable with those of systems in which the organic solvent had been included, which could be due to the efficient mass transfer between the two phases. Due to the high top to bottom phase volume ratio (8.5:1) a recovery of 98-99% could be obtained in the top phase.

After the complete transformation, the PEG rich phase could be passed over a column of hydrophobic sorbent for the extraction of prednisolone, and then recirculated to the reactor with additional substrate. The bacterial cells could be used repeatedly for the batch conversions of hydrocortisone, and also had the possibility of being reactivated by supply of nutrients.

These studies, therefore, set the basis for a continuous system for extractive bioconversion of steroids. The interesting point that comes to notice is that the combination of two techniques such as aqueous two-phase separation, and adsorption, which are generally used individually for performing extractive bioconversions, could be advantageous for efficient product recovery in case of fine chemicals as well.

Similarily, aqueous two-phase systems may find use in other areas like the extractive conversion of biosurfactants, which are assuming increased importance in biotechnology. These compounds generally affect the cells and have to be removed continuously during cultivation.

Production of macromolecules. At times, the macromolecules produced during fermentations may be toxic to the cells synthesizing them, thus preventing further synthesis. In other situations, they are so fragile, that they may be degraded in the unfavourable conditions prevailing in the fermentor. The macromolecules belonging to both the classes are generally proteinic in nature. In either case, it is desirable to remove the product as it is formed. Aqueous two-phase systems offer a promising technology to do the same. Advantage is again taken of varying volume ratios to obtain an efficient extraction of the macromolecular product into the top phase, if the partition coefficients are not favourable. Recovery into the top phase could also be enhanced by the modification of the phase polymers depending on the nature of the product.

The toxic effect of macromolecules on the bacterial cells producing them is represented very well in case of Clostridium species. The production of certain suicidal extracellular proteins (autolysins) have been reported during continuous culture of C. acetobutylicum (19), which affects the biomass concentration and also the solvent production. The toxin produced by C. tetani is also a proteolytic enzyme capable of degrading the cell wall of the bacterium; the protoplasts formed as a result, are extremely labile. In a pioneering study on the production of this toxin, Puziss and Hedén (20) showed that extractive bioconversion could be used when growing C. tetani. In an aqueous two-phase system consisting of 12% PEG 4000 and 2% Dextran (mol.wt.480 000), the toxin was equally distributed in the two phases, while the majority of the cells were in the lower phase. Because of the high top to bottom phase volume ratio (15:1) the concentration of the protein was substantially reduced in the immediate vicinity of the cells. As a result, more than 1000 fold increase in the total yield of toxin was obtained, as compared to production in conventional medium. According to the authors, the phase polymers had a stabilizing effect on the protoplasts in the older cultures, which are known to continue some metabolic activities of the cells. This could, in part, contribute to the toxin output in the phase system.

The two-phase systems have also shown potential in the production of protective antigens (20). The extractive phenomenon, thus, may be used to advantage in the production of other important cellular products such as vaccines.

The semicontinuous production of α-amylase and cellulase has been studied in PEG-Dextran systems using Bacillus subtilis and Tricoderma reesei, respectively (21, 22). Some improvements in the yields have been observed in both the cases. In case of cellulase production, the economics of the process could be improved by using a cheap, lignocellulosic waste as the substrate, and replacing the fractionated dextran with a crude one.

Optimizing the phase system is an essential prerequisite for the enzyme production, and still much work is to be done regarding this aspect. Some methods for the removal of phase polymers from the end product have been reported as described later, which have to be applied for the economical recovery of the pure enzyme.

Purification of proteins by partition in aqueous two-phase systems

Removal of cell debris, DNA and proteases, constitute the major problems faced during the large-scale isolation of intracellular proteins. This has emphasized the need for rapid prodedures for protein purification. The fundamental studies on separations in aqueous two-phase systems suggest that it is possible to spontaneously partition the protein of interest into one of the phases by selecting the suitable conditions. The method is very efficient and convenient when set up. However, it may be very laborious to find the proper conditions for a favourable partitioning. Furthermore, such a system may be influenced by variations in parameters such as composition of the cell homogenate. Several examples on large scale isolation of proteins have been reported. Some of these are listed in Table II.

Table II. Isolation of enzymes in aqueous two-phase systems.

Enzyme	Organism	Phase system	Reference
Formaldehyde dehydrogenase	Candida boidinii	PEG/crude dextran	(23)
β-Galactosidase	E.coli	PEG/salt	(24)
Fumarase	Brevibacterium ammoniagenes	PEG/salt	(25)
Aspartase	E.coli	PEG/salt	(25)
Leucine dehydrogenase	Bacillus sphaericus	PEG/crude dextran	(25)

Affinity partitioning. In some cases, it may not be possible to separate one protein out of a complex mixture by means of spontaneous partitioning in an aqueous two-phase system. Then affinity interactions may be utilized. When first described, affinity partitioning was used for purification of membrane vesicles (26), and has since been exploited in a broad spectrum of applications.

The general strategy is to design a system where all, or at least, most cell components partition to the bottom phase. By specific interaction with an affinity ligand, having a strong preference for the top phase, the protein forms an affinity complex that partitions across the phase boundary. The affinity bound material is then recovered from the top phase. In Table III are listed examples of some groups of ligands used in this context.

The ligands are often modified by coupling PEG to them. The coupling chemistry when dealing with PEG-modification is the same as in conventional affinity chromatography. Usually, a conjugate between PEG and one ligand molecule is formed. The lifting power of such a ligand may very well be enough when purifying multivalent structures, e.g. membrane vesicles or oligomeric proteins. In other cases, e.g. when the ligand is a macromolecule that _per se_ prefers the bottom phase, then a higher degree of modification is needed. Lectins and antibodies used for selective extraction of glycosylated structures and antigens respectively, are amongst the ligands belonging to this category.

Table III. Examples of affinity partitioning in aqueous two-phase systems.

Ligand	Compound purified	Source	Reference
Cibacron blue	Formate dehydrogenase	_Candida boidinii_	(27)
Concanavalin A	Peroxidase	Horseradish	(28)
Antibodies	β_2-microglobulin	Human plasma	(29)
	Erythrocytes	Human blood	(30)
Coenzyme (NADH)	Dehydrogenases	_Candida boidinii_	(27)

However, a problem encountered when using heavy modification is that the binding capacity may go down. This can be attributed to a chemical modification of groups essential for the affinity interaction, or may be due to sterical hindrance by the polymer chains bound to the ligand. If this decrease is severe, the whole procedure may be useless unless a second separator is used as a modified group that biospecifically binds the ligand which, in turn, interacts with the structure to be isolated. Examples of such secondary separators are cells of _Staphylococcus aureus_ carrying protein A on their surfaces that binds to the F_c part of the immunoglobulin G, leaving the F_{ab} part free for the binding of antigen. _S. aureus_ cells can be heavily modified by methoxy polyethylene glycol (MPEG) to make them partition to the top phase and still maintain their binding capacity for IgG-molecules. Another such separator is avidin that is MPEG-modified, retaining its binding capacity for biotin. Hence, a general scheme for partitioning of molecules of interest to the top phase, is obtained if they are labelled with biotin. In Table IV are listed some data on partition behaviour of the above discussed secondary separators.

Aqueous two-phase systems are rather subtle. The sensitivity of such systems is very well demonstrated by Veide _et al._ (31) in that a mixture of PEG and potassium phosphate giving a homogeneous solution, suddenly started to form a two-phase system when cell homogenate was loaded in the system. Minor variation in ionic composition may also give dramatic changes in partition behaviour. This may be regarded as an advantage when dealing with well known, stable systems, whereas in other cases, it may lead to unpredictable behaviour. Hence, it may be desirable to design the affinity partitioning in such a way that the partition behaviour of the affinity complex can be predicted, and is stable within a wide range of external conditions. The approach described for _Staphylococcus aureus_ above represents such a robust predictable partition behaviour. A stable system is easier to operate, but may prove quite expensive, thus limiting its application to analytical purposes.

Table IV. Distribution in two-phase systems of separators, Staphylococcus aureus and avidin.

	Staphylococcus aureus		Avidin
	Native	MPEG-modified	MPEG-modified
Top phase	0 %	80 %	90 %
Interface	10 %	20 %	0 %
Bottom phase	90 %	0 %	10 %

Note: A phase system consisting of 0.15 g/g PEG-4000 and 0.15 g/g $MgSO_4 \cdot 7H_2O$ was used.
Source: Reproduced with permission from Ref. 29. Copyright 1983, Elsevier Biomedical Press.

Recently, another variation on this theme was presented. By introducing PEG-modified affinity sorbents (Sepharose beads carrying the immobilized ligand) a new process configuration for protein purification was achieved as shown in Figure 2 (32). The beads were exposed to the cell homogenate before the phase components were added. After phase separation the particles were recovered from the top phase, as a layer on top of the interface. The beads were collected and transferred to a column, where they were washed free of the phase polymers, and then the elution was carried out according to conventional procedures.

By this technique of combining affinity partitioning with affinity chromatographic elution, the advantages of the two procedures were combined. The rapid and effective removal of cell debris in the extraction procedure, and finally, the efficient elution procedure of the affinity chromatographic step was achieved.

Extraction in aqueous two-phase systems is said to be an operation that is easy to scale up (33). This is also true for affinity partitioning and is clearly illustrated by the process for purification of formate dehydrogenase (34, 35). In this study, small scale experiments gave an overall enzyme yield of 74%. When scaled up by a factor of 40 000, the yield was 70%, thus demonstrating the feasibility of evaluating the performance of extraction process on a small scale.

Removal of polymers

A general problem that is specific for the two-phase extraction technique is the removal of polymers from the purified protein. In the case of PEG-salt systems, a general strategy has been to extract one protein to the PEG-rich phase and then replace the salt phase by a fresh salt phase under conditions in which the protein prefers the salt phase. This phase is then separated and the salt removed by

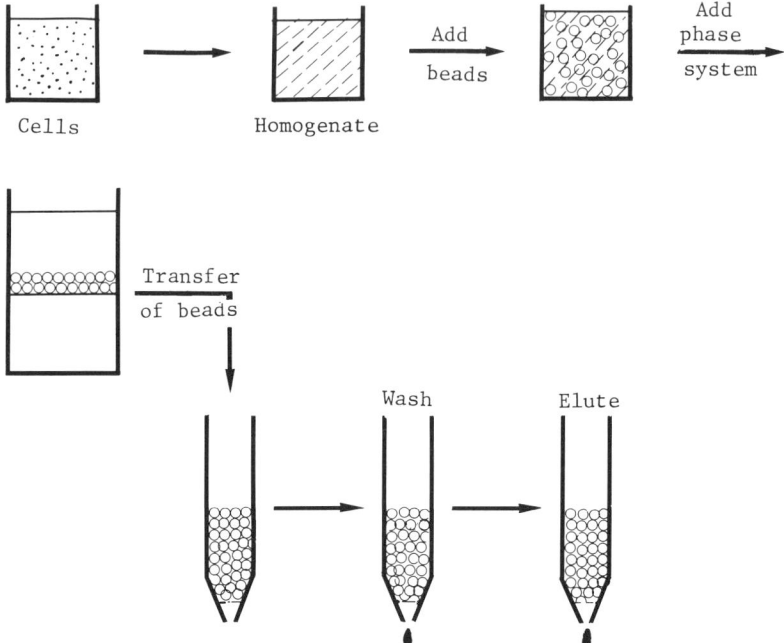

Figure 2. Partitioning in two-phase system with the use of an affinity sorbent. Reproduced with permission from Ref. 2. Copyright 1986, Wiley-Interscience.

ultrafiltration or dialysis (35). However, one has to take into consideration the fact that there will also be some PEG in the salt phase and that this amount in relation to the amount of protein is quite large. Low molecular weight PEG can be removed in the dialysis step. Alternative methods to remove the phase components have been presented. Ion exchange (2) or hydrophobic chromatography (37) gives opportunities to adsorb the charged proteins and let the uncharged polymers pass. The use of particle bound ligands mentioned above enables one to wash off the polymers prior to elution of the affinity-bound material.

Economy

A hampering factor in the exploitation of aqueous two-phase systems in biotechnology has been the cost of the phase components. The cost of PEG/salt systems have been regarded as acceptable, whereas polymer/polymer systems, usually consisting of fractionated dextran as bottom phase polymer, have been too expensive to use in large scale applications. Efforts to use crude dextran led to a reduction in cost, but not to an extent that made it worthwhile to pursue. The crude dextran produced by Leuconostoc mesenteroides is too viscous to be used directly. Partial hydrolysis had to be applied and that increased the cost substantially.

A reason why the PEG/salt systems have not been used more extensively is that the high ionic strength severely influences the affinity interactions when affinity partitioning is to be exploited. However, in applications where spontaneous partitioning is sufficient, these systems are often used. Recently, the use of starch derivatives named as Reppal PES has been reported (32) that are useful in forming aqueous two-phase systems. The properties of these new polymers resemble those of dextran.

Based on the costs for the polymers of 2.5 $/kg and 7 $/kg for PEG and Reppal, respectively, one can calculate the cost of the phase system. For two equivalent phase systems having dextran and Reppal as bottom phase polymers respectively, the cost is reduced from 5.5 $/L for PEG/dextran to 0.50 $/L for PEG/Reppal.

A crucial point when discussing economics of phase systems is the loading of the system. The more biological material that can be loaded, lesser is the cost of phase system per unit processed. Furthermore, extensive studies have been carried out concerning recirculation of the phase components. The latter efforts have mainly been focused on PEG/salt systems, and also on the polymer/polymer phase systems employing affinity ligands. Kula and coworkers reported the reuse of the salt phase when carrying out extractive purification (37). They showed that the phases could be recycled 4 times without any marked loss in performance of the phase separation and in specific activity of the purified protein.

An additional aspect of the phase components to be considered is their polluting effects. If the used phases are pumped down the drain, then the cost of the waste water treatment plant has to be included in the process economics. Phosphate is regarded as one of the more difficult nutrients to be removed efficiently in waste water treatment process. Biodegradable polymers seem much more

acceptable from the pollution point of view. PEG is biodegradable, though by a slow process. Recycling of phase components will thus reduce the direct costs and also the loading on the recipient systems.
The systems are economically acceptable when separating proteins out of a cell homogenate. On the other hand, when carrying out bioconversion processes, the phase system has to be continuously reused, hence, there must be a method to remove the products from the phase system. This can be performed in a number of ways like adsorption, membrane filtration, etc.

Equipment for continuous extraction

It has been demonstrated that if an efficient separation of different substances in an aqueous two-phase system, cannot be achieved in a single step, a multistage procedure could be resorted to (2). This method, well-known as counter current distribution (CCD), involves large number of extraction steps in order to separate substances differing only slightly in partition coefficients. Thus, it has been possible to fractionate red blood cells based on their age (38,39); it has also been demonstrated that the wing buds in chick embryos, at a very early stage of development contain cell populations with different surface properties, and also, later during the development, show structural differences (39, 40). Another illustrative example of the extreme resolving power in such multistage extractive processes is the demonstration that enzymes that operate in sequence in the cell but so far, have never been shown to express any affinity for each other, partition together in the phases (41). If this high resolving power were exploited, a very efficient purification process would be set up.

The CCD machines are characterized by the presence of a series of compartments containing defined volumes of top and bottom phases, in which a substance partitions itself depending on its partition coefficient. The phase mixing is followed by the settlement of the phases, and moving the top phases to the adjacent bottom phases. This process is repeated depending on the number of compartments in the machine. Normally, a CCD machine has a low capacity, despite a high resolving power (2).

A centrifugal step can be introduced in the process to reduce the phase settling time. This has been made possible by the use of counter-current continuous flow centrifugal separators developed by Sanki Engineering Ltd. (Japan), which also enable the CCD to be driven continuously, with high throughputs. In these continuous processes it is, of course, suitable also to introduce affinity interactions.

Conclusion

The versatility and the potential of aqueous two-phase systems in future biotechnology has been amply demonstrated. The applications described here deal with extractive bioconversions, isolation and purification of proteins. Biochemical analyses in terms of binding assays have also been successfully applied in the two-phase systems (1).

The biocompatibility, fast separation process and the ease to scale up are some of the characteristics that make aqueous two-phase systems an attractive alternative to other known techniques when biochemical separation is to be carried out.

Acknowledgments

The authors are grateful to the National Swedish Board for Technical Development for financial support.

Literature cited

1. Mattiasson, B. Trends Biotechnol. 1983, 1, 16-20.
2. Albertsson, P.-Å. "Partition of Cell Particles and Macromolecules"; 3rd ed., Wiley-Interscience: New York, 1986.
3. Mattiasson, B.; Hahn-Hägerdal, B. In "Immobilized Cells and Organelles"; Mattiasson, B., Ed.; CRC: Florida, 1983; Vol. I, p. 121.
4. Monsan, P.; Combes, D. Ann. NY Acad. Sci. 1984, 434, 48-60.
5. Tjerneld, F.; Johansson, G.; Berner, S.; Persson, L. 4th Int. Conf. Partition in Aqueous Two-Phase Systems, Lund, Aug 1985, Abstract p 25.
6. Ling, T.G.I.; Mattiasson, B. 4th Int. Conf. Partition in Aqueous Two-Phase Systems, Lund, Aug 1985, Abstract p. 45
7. Larsson, M.; Mattiasson, B. Chemist.Indust. 1984, 12, 428-431.
8. Hahn-Hägerdal, B.; Larsson, M.; Mattiasson, B. Biotechnol. Bioeng. Symp. No. 12, 1982, 199-202.
9. Mattiasson, B. In "Immobilized Cells and Organelles"; Mattiasson, B., Ed.; CRC: Florida, 1983; Vol. II, p. 23.
10. Hahn-Hägerdal, B.; Mattiasson, B.; Albertsson, P.-Å. Biotechnol. Lett. 1981, 3, 53-58.
11. Mattiasson, B.; Larsson, M. Proc. 3rd Eur. Cong. Biotechnol. Verlag Chemie: Weinheim, 1985, Vol. IV, p. 375.
12. Kuhn, I. Biotechnol. Bioeng. 1980, 22, 2393-2398.
13. Mattiasson, B.; Suominen, M.; Andersson, E.; Häggström, L; Albertsson, P.-Å.; Hahn-Hägerdal, B. In "Enzyme Engineering"; Chibata, I.; Fukui, S.; Wingard, L.B., Jr., Eds.; Plenum: New York, 1982, Vol. 6, p. 153.
14. Mattiasson, B.; Larsson, M. In "Biotechnology and Genetic Engineering Reviews"; Russel, G.E., Ed.; Intercept:England, 1985, Vol. 3, p. 137.
15. Andersson, E.; Mattiasson, B.; Hahn-Hägerdal, B. Enzyme Microb. Technol. 1984, 6, 301-306.
16. Proc. Symp. on Biocatalysts in Organic Syntheses; Tramper,J.; Van Der Plas, H.C.; Linko, P., Eds.; Elsevier: Amsterdam, 1985.
17. Antonini, E.; Carrea, G.; Cremonesi, P. Enzyme Microb. Technol. 1981, 3, 291-296.
18. Kaul, R.; Mattiasson, B., unpublished data.
19. Soucaille, P.; Minier, M.; Ferras, E.; Goma, G. Preprint 3rd Eur. Cong. Biotechnol., Munich, Sept 1984, Vol. II, p. 85.

20. Puziss, M.; Hedén, C.-G. Biotechnol. Bioeng. 1965, 7, 355-366.
21. Andersson, E.; Johansson, A.-C.; Hahn-Hägerdal, B. Ann. NY Acad. Sci. 1984, 434, 115-118.
22. Persson, I.; Tjerneld, F.; Hahn-Hägerdal, B. 4th Int. Conf. on Partition in Aqueous Two-Phase Systems, Lund, Aug 1985, Abstract p. 20.
23. Kroner, K.M.; Hustedt, H.; Kula, M.-R. Biotechnol. Bioeng. 1982, 24, 1015-1045.
24. Veide, A.; Smeds, A.-L.; Enfors, S.-O. Biotechnol. Bioeng. 1983, 25, 1789-1800.
25. Hustedt, H.; Kroner, K.H.; Menge, U.; Kula, M.-R. Trends Biotechnol. 1985, 3, 139-144.
26. Flanagan, S.D.; Barondes, S.H. J. Biol. Chem. 1975, 250, 1484-1489.
27. Kula, M.-R.; Johansson, G.; Buckmann, A.F. Biochem. Soc. Transac. 1979, 7, 1-5.
28. Mattiasson, B.; Ling, T.G.I. J. Immunol. Methods 1980, 38, 217-223.
29. Ling, T.G.I.; Mattiasson, B. J. Immunol. Methods 1983, 59, 327-337.
30. Sharp, K.A.; Yalpani, M.; Howard, S.J.; Brooks, D.E. 4th Int. Conf. Partition in Aqueous Two-Phase Systems, Lund, Aug 1985, Abstract p. 38.
31. Veide, A.; Lindback, T.; Enfors, S.-O. Enzyme Microb.Technol. 1984, 6, 325-330.
32. Mattiasson, B.; Ling, T.G.I. J. Chromat. (in press).
33. Kula, M.-R.; Buckmann, A.; Hustedt, H.; Kroner, K.H.; Morr, M. In "Enzyme Engineering"; Broun, E.B.; Manecke, G.; Wingard, L.B.,Jr., Eds; Plenum: New York, 1978, Vol. 4, p. 47.
34. Kroner, K.H.; Schütte, H.; Starch, W.; Kula, M.-R. J. Chem. Tech. Biotechnol. 1982, 32, 130-137.
35. Cordes, A.; Flossdorf, J.; Kula, M.-R. Preprint 3rd Eur. Cong. Biotechnol. Munich, Sept 1984, Vol. III, p.557.
36. Hustedt, H.; Kroner, K.H.; Menge, H.; Kula, M.-R. Preprint 1st Eur. Cong. Biotechnol., Interlaken, Sept 1978, Part 1, p. 48.
37. Kula, M.-R. Lecture presented at 4th Int. Conf. Partition in Aqueous Two-Phase Systems, Lund, Aug 1985.
38. Martin, M.; Tejero, C.; Galvez, M.; Pinilla, M.; Luque, J. Acta Biol. Med. Ger. 1981, 40, 979-984.
39. Walter, H.; Pangburn, M.K. 4th Int. Conf. Partition in Aqueous Two-Phase Systems, Lund, Aug 1985, Abstract p.4.
40. Sharpe, P.T.; Cottrill, C.P.; Wolpert, L. 4th Int. Conf.Partition in Aqueous Two-Phase Systems, Lund, Aug 1985, Abstract p. 7.
41. Bachman, L.; Johansson, G. FEBS Lett. 1976, 65, 34-43.

RECEIVED March 26, 1986

8

Recovery of Proteins from Polyethylene Glycol–Water Solution by Salt Partition

G. B. Dove and G. Mitra

Cutter Laboratories, Berkeley, CA 94710

>   Addition of salts (e.g. potassium phosphate dibasic) partitions an aqueous system containing 20% w/v polyethylene glycol (PEG) into two liquid phases: a PEG enriched phase and a salt enriched phase. Proteins and polymers (e.g. DNA, albumin, immunoglobulins, alpha-1 antitrypsin, and PEG) distribute unevenly between the two phases. Partition coefficients (concentration in PEG phase / concentration in salt phase = K) are influenced by physical parameters, such as salt composition and concentration, pH (ion ratios) and temperature. Specific proteins (e.g. alpha-1 antitrypsin) exhibit low K values in a wide range of conditions.
>   Higher salt concentrations and pH yield higher partition coefficients. In a plasma source, the K of alpha-1 antitrypsin is 0.0006 at 0.5 M salt and increases to 0.0062 at 1.6 M salt. The K of PEG increases to 200+ in 1.0 M salt. Proteins in general exhibit K values of 0.01-100. Altering pH to make proteins or other partitioned materials more/less hydrophilic induces greater/lower solubility. A pH change from 5 to 9 increases K in general by 100-fold+. Further, the trends demonstrated by increasing salt concentration are amplified. Lower temperatures (5 to -5 C) increase the PEG K by two to ten-fold with little change in protein distribution. Conditions may be tailored to optimize isolation of specific proteins to permit recoveries of 90% from mixed systems, such as plasma or fermentation broths.

This paper is organized into three parts. Purification techniques are outlined briefly in comparison to aqueous extraction, followed by a review of properties and work in multiphase systems with emphasis on the purification of proteins. Finally, recent work undertaken by the authors is presented, involving proteins of pharmaceutical importance.

Various methods are available for the separation of biochemicals. These include physical methods of centrifugation and filtration, chemical methods of precipitation and extraction, and interactive techniques, such as electrophoresis and chromatography. These methods are employed to perform the steps necessary to purify biological materials from complex solutions. The isolation of a specific component (e.g. a protein) from a plasma source or a fermentation broth requires several steps:
a) removal of cell particles (disruption if intracellular product), e.g. centrifugation.
b) preliminary purification, e.g concentration, precipitation.
c) secondary purification, e.g. high-resolution chromatography.
d) finishing.

Through these steps, the necessary purity and yield are achieved. Requirements for purity and yield are dictated by final use. Purity may range as low as 10% for bulk enzymes to virtually 100% for therapeutic use. The final yield affects directly the cost of the finished product and is of critical economic importance, as feed stocks for the processes are typically expensive. Final yields are considered in terms of biologically active material, as many of the components are labile and useless in a denatured state. Specifically, the constraints of high purity and biologically-active yield in the production of therapeutic products limit the alternatives available for purification processes.

Extractions and precipitations in chemistry are well-established for organic systems. For example, nucleic acids may be extracted in a phenol/water mixture ($\underline{1}$). The use of aqueous extractions has several advantages over these and other widely used methods of separation. 1) Chemical components, such as polymers and/or salts, may be chosen to minimize denaturation due to solvency or interfacial tension ($\underline{2}$). Solvent/water mixtures, such as phenol/water, produce interfacial tensions in the range of 50 dyne/cm, compared to 0.1 dyne/cm in aqueous systems($\underline{3}$). 2) Physical sources of denaturation are minimal, with virtually no shear. Simple mixing only is required; centrifugation may be used to hasten separation ($\underline{4}$). 3) Conditions may be tailored to satisfy specific isolation requirements by basing separations on dissimilar solubilities and affinities, which are dependent on pH and salts. These effects are not applicable to other methods. 4) The process is easily scaled to any volume of material, with minimal capital investment($\underline{4}$). Beyond conventional products, the technique is applicable to biotechnical separations with unique possibilities, which will be addressed below.

Properties and Applications of Aqueous Systems

The methodology of aqueous extractions is adaptable to the requirements of isolating biologics due to the high water content of the system. The addition of water-soluble polymers and/or salts to water produces spontaneously two or more liquid phases. The denser solution (usually the salt-rich one) forms the bottom phase. Each phase is comprised primarily of water (80-95%)($\underline{2}$).

The top and bottom liquid regions are separated by an interface (P1). Precipitates may form at this liquid-liquid interface or may settle to the bottom of the vessel (P2), forming a solid-liquid interface. Any material in the mixture may distribute to any region, dependent on the combined properties of all components.

Multiphase liquid systems are analogous to precipitations in that precipitations consist of one liquid phase and one solid region, whereas multiphase liquid systems possess two or more liquid regions and a range of solid regions. The increased number of environments available to a material make isolation more amenable to optimization.

In a given system of a polymer, a salt and water, two liquid phases exist with the corresponding equilibrium concentrations of each component. Typically, the top phase is enriched with polymer and the bottom phase is enriched with salt. The phases partition because of mutual "incompatibility" (2). A protein in the system will distribute between the two phases according to the properties of the partitioning agents and other materials present. A partition coefficient (K) may be defined as (2):

$$K = Ct/Cb \qquad (1)$$

where Ct and Cb are the concentrations of protein in the top and bottom phases, respectively. If a component has an equal affinity for both phases, the concentrations in each phase are equal and K is equal to unity.

The partition coefficient of any component in the system (polymer, salt, protein) may be manipulated by several parameters (4):
a) Chemical basis and molecular weight of primary polymer (e.g. polyethylene glycol: PEG).
b) Chemical basis of second agent (e.g. polymer, salt).
c) Concentrations of additives (e.g. proteins).
d) pH (or ionic ratios).
e) Temperature.

The number of controllable physical parameters is large. Parameters may be varied individually in methodical fashion, as was done in this work. Experiments necessary to optimize a system for a specific component may be reduced by the simplex method (5) or fractional factorial design (6).

In the last thirty years, substantial data has been accumulated for aqueous multiphase systems. The data may be classified by the components necessary to form multiple phases. Non-ionic polymers, polyelectrolytes, low molecular weight organics and salts have been studied to establish phase diagrams for two or more components in defined systems (Albertsson, 2). The most common systems investigated have been mixtures of PEG and dextran or a phosphate salt. Separations by the simple partitioning agents can be enhanced by changes in the chemical and physical environment. The addition of minor contaminants (e.g. detergents or other surfactants) alters the surface tension (7). Modification of polymers, such as attaching a charged group to PEG (8) or dextran (9), and applications to affinity ligand systems (9, 10) and assays (11) are under investigation. Mass transport across the interface has been studied (12).

Partitioning of plasma proteins (2, 6, 13), enzymes (see below), cells and cell particles (2, 14, 15) and nucleic acids (16, 17) in a variety of systems have been reported.

Large scale purification of enzymes in PEG/dextran and PEG/salt have been reported and reviewed by Kula et. al.(18, 19, 20). Examples include pullulanase and 1,4-a-glucan phosphorylase (21), a-amylase (22), b-galactosidase (23), and the general economics of extractive enzyme recovery (24). Process studies have covered the use of continuous centrifuges (18, 19) and countercurrent distribution trains (2, 25).

Purification of b-galactosidase from E. coli may be compared: Higgins (26) outlines a process of succeeding centrifugations to remove cell debris, nucleic acids precipitate, and protein precipitate (product). Veide (23) outlines a single aqueous extraction with PEG and salt in which b-galactosidase partitions to the PEG-rich phase. Cells, nucleic acids, and a major part of the contaminating proteins partition to the salt-rich phase.

Several applications in process purification serve to illustrate the diversity of uses. The partitioning of phases allows components to be separated within the confines of another operation. Production of a material (e.g. fermentation and subsequent cell separation) may be simplified. Tissue culture broth with cells may be sonicated. The cell walls are disrupted to release intracellular products and degrading enzymes. The broth is partitioned quickly with PEG/salt in bulk.

In an unmodified PEG/dextran system, cells and cellular components partition to the dextran-rich bottom phase (2), leaving the top phase available for product. a-amylase has been partitioned to the top phase and Bacillus subtilis cells to the bottom phase (22). Production of biologically active materials may be enhanced by removal of product from cells or cellular components for two reasons: 1) Degradation of product by extra- or intra-cellular enzymes still present in the broth is prevented. 2) Removal of product reduces negative feedback inhibition of growth or production of cells. The method is unique in speed and ease for handling bulk quantities which is critical for sensitive systems.

A different approach involves the use of ligands covalently attached to PEG, as noted above (8). The ligand/PEG complex is introduced to the broth, where the ligand binds its target complement, a product. PEG and salt are added; the PEG/ligand/-complement complex partitions preferentially to the PEG phase. The complement has been isolated in the PEG phase, whereas its intrinsic distribution would favor the salt phase. The separation/purification is high where conditions are such that all other materials partition to the salt phase.

This study was initiated to explore the applications of partitioned phases to separations of therapeutically active materials. Posssibilities include: 1) isolation and purification of one component from a complex mixture (e.g. alpha-1 antitrypsin, immunoglobulins from plasma sources or tissue culture broth), 2) removal of a secondary product or contaminant (e.g. DNA, residual PEG from a process stream) and 3) simultaneous fermentation and isolation of product from cellular components.

A system of PEG/salt was chosen because the settling time is short (10-60 min.) compared to polymer/polymer systems (PEG/dextran is 30-180 min.). Typically, PEG levels under similar conditions are lower in salt systems. Further, it was observed that PEG levels could be substantially reduced in the salt phase by optimization. The data presented in this paper represent part of the work in progress.

Materials/Methods

Polyethylene glycol (PEG 3350), HO(CH2-CH2O)x-CH2-CH2OH, was obtained from Union Carbide. PEG 3350 has a molecular weight of 3300-3400. Reagents (potassium phosphate dibasic, phosphoric acid) were obtained from J.T. Baker, "Baker analyzed" reagent grade.

Simple systems (with a single defined additive) were produced with each of the following materials. Calf thymus DNA, polymerized, was obtained from Sigma. Protein sources were prepared in-house and subsequently dialyzed into low salt solutions. Human serum albumin and immunoglobulin G (IgG) were plasma-derived. Human immunoglobulin M (IgM) was produced by tissue culture fermentation and purified. A defined complex system consisted of both albumin and IgM together. An undefined complex system was set up with an intermediate material of Cohn plasma fractionation containing alpha-1 antitrypsin (alpha-1), albumin, and other contaminants.

A stock solution of 40% PEG was stored at 5 C. Simple systems were formed by the addition of PEG solution, salt, and water to give 20% w/v PEG and appropriate salt. Solutions were mixed and adjusted to pH with phosphoric acid. DNA was added to an approximate concentration of 1 mg/ml. Protein concentrates in unbuffered solutions were added to an approximate concentration of 1-10 mg/ml. To the plasma fraction, PEG and salt crystals were added. Systems were gently mixed by rocking in polypropylene centrifuge tubes. The mixtures were allowed to settle overnight at -4, 5 or 20 C. Tubes were centrifuged at 2000 RCF for 30 min.

Samples were assayed by absorbance at 280 nm, Bradford protein assay(27), and radial immunodiffusion plates (Helena Laboratories). Alpha-1 antitrypsin was assayed for biological activity by competitive assay with elastase (28). DNA was assayed by the diphenylamine methods of Burton(29) and Giles and Myers(30), with modifications due to the presence of PEG and salts (31). Materials were also assayed by size exclusion chromatography on Superose 6 (Pharmacia FPLC), with peak integration at 280 nm. PEG was assayed by HPLC.

Results

Data are presented in several forms for many of the partitioned materials. The concentration of material in the salt phase and the partition coefficient (concentration in the PEG phase / concentration in the salt phase = K) are plotted as functions of salt concentration and pH on semi-log scale. On the log axis, 1E0 represents $1 \times 10(0)$ or 1; 2E3 represents $2 \times 10(3)$ or 2,000. A K value of 1 indicates no preference for either phase; the concentration is the same in both. It is this value of unity that is critical. Values greater than 1 indicate a preference for the top (PEG) phase and values less than 1 indicate the bottom (salt) phase. In several cases, precipitation of protein occurs at the interface or at the bottom of the tube. As precipitation occurs, the concentration of protein in the salt phase may decrease without a concurrent increase in the partition coefficient.

Centrifugation after settling for 24 hours had no effect on the volume ratios (PEG phase / salt phase). Gelatinous precipitates were observed in systems at higher salt and lower pH due to saturation of the system with respect to the protein at the given salt/pH environment. Centrifugation compressed this precipitate layer to less than 10% of the phase volume.

In general, the volume ratio (PEG phase / salt phase) decreases as the salt concentration or pH increases.

Table I. Relative volumes of phases as a function of salt concentration in a typical system at 5 C.

Salt Concentration (M)	PEG (% vol.)	Salt (% vol.)
0.5	78	22
0.6	61	39
0.8	48	52
0.8 (-5 C)	47	53
1.0	44	56
1.2	44	56
1.6	44	56

The gross physical characteristics of the system can be influenced by relatively minor changes in composition. For example, adjustment of pH with hydrochloric acid instead of phosphoric prevents an interface from forming below pH 5.5.

DNA, IgM, IgG, albumin, and alpha-1 antitrypsin follow similar trends (Table II, Figures 1-4). As the salt concentration increases, the concentration of material in the salt phase decreases and the partition coefficient increases. IgG exhibits the same pattern with increasing pH (Figure 5). Other materials are not as consistent. The trends appear to be somewhat additive: increasing the salt concentration and pH lead to the greatest partition coefficients. Temperature changes give mixed results with little change (Figures 2, 4, 5). In several instances, systems at the extreme values of the parameter ranges indicate an amplification of trends observed in the middle ranges.

Various forms of DNA behave differently. In PEG/dextran systems, native DNA exhibits higher values of K than denatured DNA. Both materials exhibit higher K values as the pH is increased (i.e. $(H2PO4)-$ is shifted to $(PO4)3-$) (16).

A summation of partition coefficients at pH 9 is given in Figure 6. All materials exhibit a tendency toward the salt phase at low salt concentrations. As the salt concentration increases, DNA, albumin, IgM, and IgG are repelled from the salt phase to the PEG phase. Alpha-1 remains in the salt phase exclusively. The pattern at pH 9 stands in contrast to the results at pH 5, in Figure 7. Materials do not migrate to the PEG phase. K values remain below 1 even in high salt concentrations.

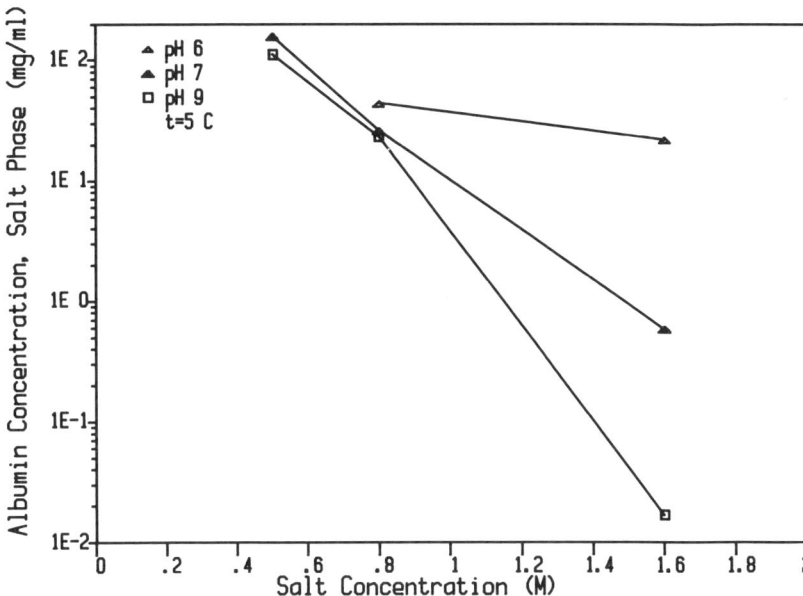

Figure 1. Albumin concentration in the salt phase as a function of salt concentration.

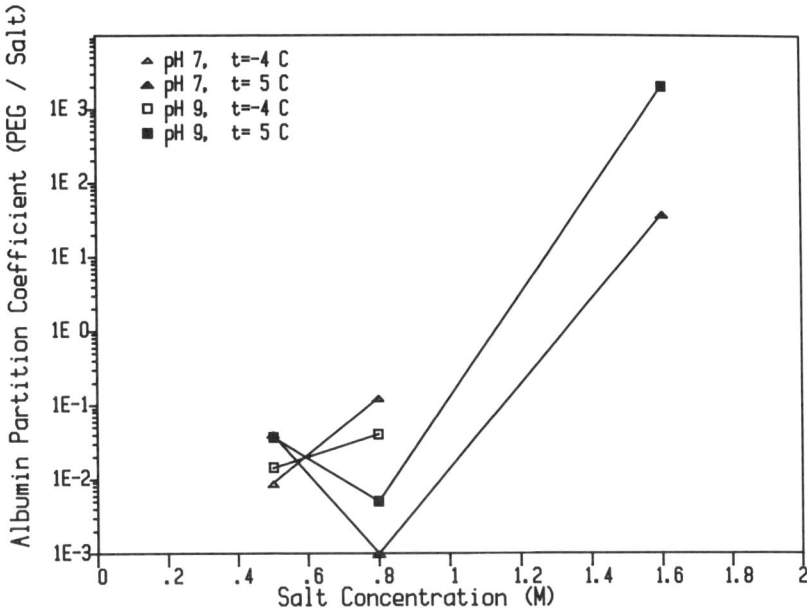

Figure 2. Albumin partition coefficient (concentration in PEG phase / concentration in salt phase) as a function of salt concentration.

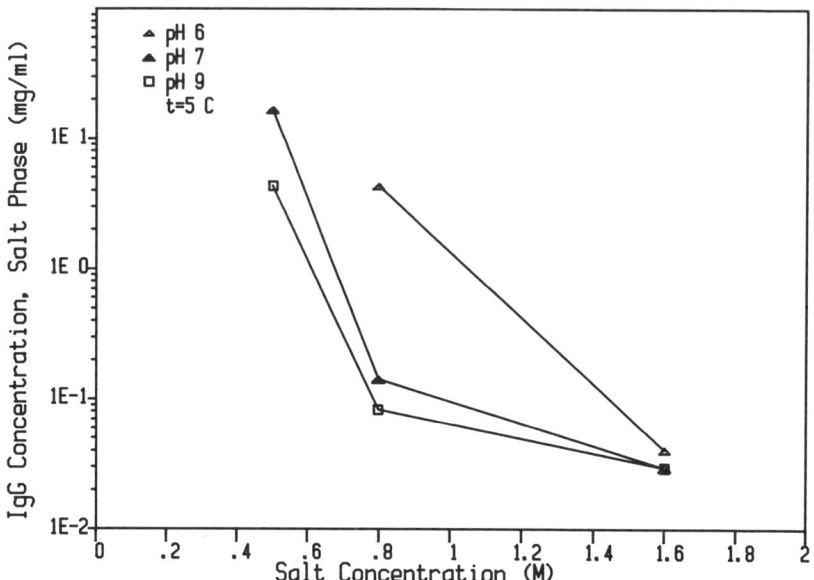

Figure 3. Immunoglobulin G concentration in the salt phase as a function of salt concentration.

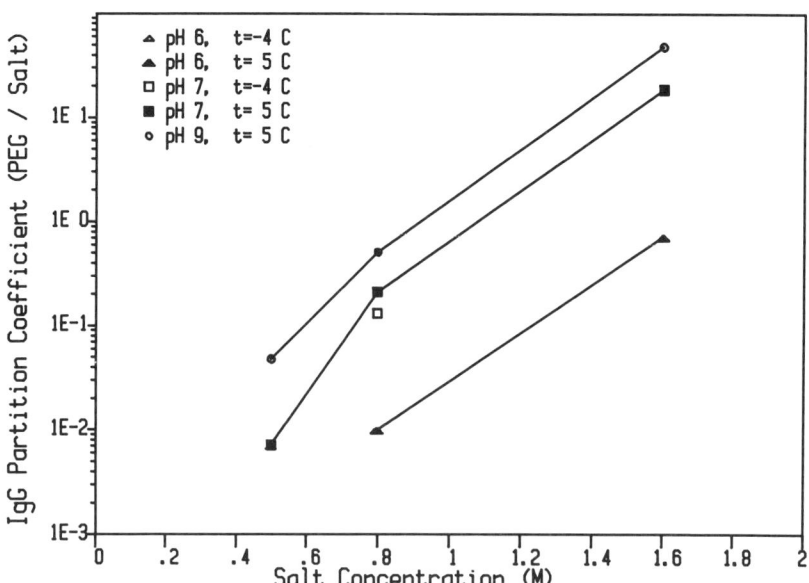

Figure 4. Immunoglobulin G partition coefficient as a function of salt concentration.

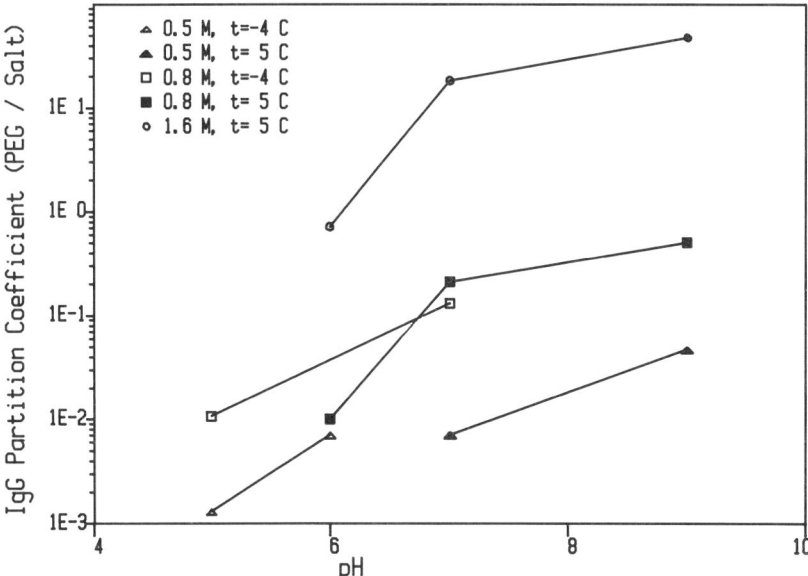

Figure 5. Immunoglobulin G partition coefficient as a function of pH.

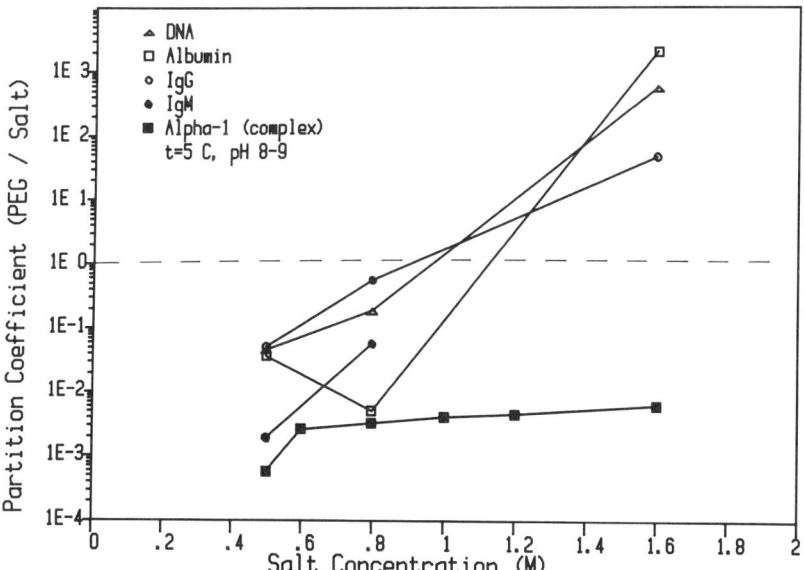

Figure 6. Summary of partition coefficients at pH 8-9 as a function of salt concentration.

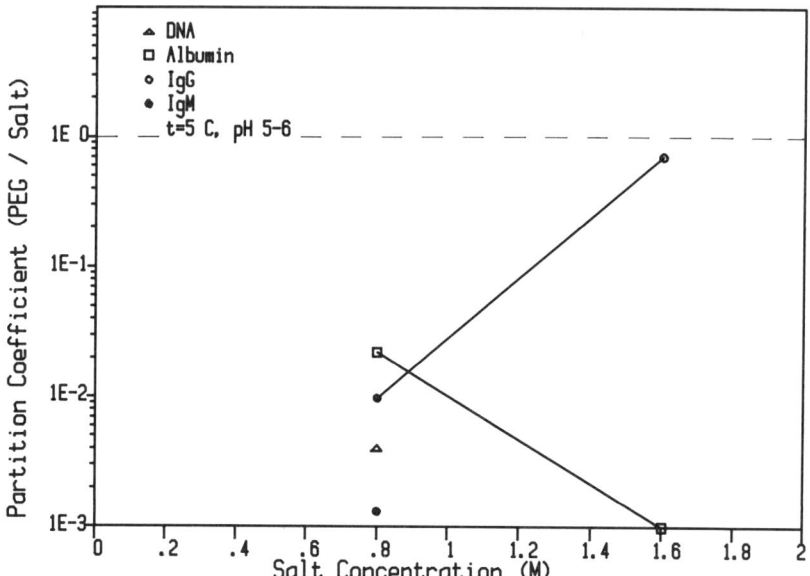

Figure 7. Summary of partition coefficients at pH 5-6 as a function of salt concentration.

Table II. Concentration of materials in the salt phase and partition cefficient as functions of salt concentration and pH.

Material t= 5 C	Salt Conc'n (M)	pH	Conc'n (Salt phase)	Partition Coeff. (PEG/Salt)
DNA			(ug/ml)	
	0.5	9	23.5	0.043
	0.8	6	337	0.004
	0.8	7	5.6	0.179
	0.8	9	5.4	0.185
	1.6	9	1	538
IgM			(mg/ml)	
	0.5	9	1.0	0.0019
	0.8	6	1.4	0.0013
	0.8	7	0.080	0.024
	0.8	9	0.035	0.054
	1.6	9	0.0019	n.a.
Alpha-1			(mg/ml)	
	0.5	8	17.8	0.00056
	0.6	8	10.1	0.0025
	0.8	8	6.3	0.0033
	1.0	8	5.5	0.0038
	1.2	8	4.7	0.0044
	1.6	8	3.3	0.0062

The difference between alpha-1 and other proteins may be attributed to either the intrinsic nature of alpha-1 (e.g. low hydrophobicity) or the presence of miscellaneous plasma fraction contaminants in the salt phase attracting alpha-1 or in the PEG phase repelling alpha-1. A mixture of albumin and IgM show qualitatively the same values as the respective simple systems. As the salt concentration increases, the concentration of protein in the salt phase decreases and the partition coefficient increases. Again, at lower pH, materials do not migrate to the PEG phase and partition coefficients do not exceed 1. Further work is in progress to define the separation between multiple components based on experiments with defined systems.

The effects of salt concentration and pH have been studied previously (2, 16, 32). It has been found that the concentration and pH are not as critical as the ratio of ions. The pH is a measure of the ionic environment; that is, the ratio of charged ions derived from the salt, K2HPO4 and acid, H3PO4. Compared to effects of small ions, the concentration of proteins has little effect on partition coefficients (32). In general, higher valent anions yield higher partition coefficients: K of (PO4)3- > (HPO4)2- > (H2PO4)-. The effects of cations and anions are cumulative. Ionic effects can include the addition of NaCl to PEG/dextran systems (2).

The distribution may be defined in terms of a model equation. From the Bronsted formula (2, 33, 34):

$$K = e^{(M x)/(R T)} \qquad (2)$$

where M is the molecular weight of the partitioned component, x is a factor dependent on the component and system other than size, R is the gas constant and T is the temperature. To predict qualitatively the behavior of the system, existing data may be extrapolated. Calculating x with IgG data and extrapolating it to IgM, the values of Table III are generated.

Table III. Calculation of Bronsted formula parameters with IgG data and extrapolation to IgM at 5 C.

Salt Conc'n (M)	0.5 M	0.8 M	0.8 M	0.8 M
pH	9	6	7	9
IgG (K, data)	5E-2	1E-2	2E-1	6E-1
IgG (x)	-4.3E-4	-6.6E-4	-2.3E-4	-7.3E-5
IgM (K, calc.)	3E-7	1E-10	3E-4	8E-2
IgM (K, data)	2E-3	1E-3	2E-2	7E-2

For IgM, K (calc.) / K (data) is less than 1. The values converge only at higher salt and pH. Data indicate a higher affinity for the PEG phase than predicted by extrapolation of IgG values in the Bronsted formula. If the PEG phase tends to attract, or the salt phase repels, hydrophobic species, then IgM exhibits greater hydrophobicity than expected. Conversely, calculations to predict the behavior of IgG based on IgM would indicate greater hydrophilicity. It may be hypothesized that the globular nature of IgM shields the Fc terminal carboxyl group from bulk interactions, reducing the net charge density of the molecule in solution.

A change in temperature of 15 C yields a change in calculated values of K of less than 5%. Data indicate no significant change in protein K values over the observed temperature range.

Applications

The practical consequences of a process separation involving alpha-1 antitrypsin are indicated in Figures 8 and 9. The yield (Figure 8) is the product of the concentration of alpha-1 in the salt phase and the volume of the phase. Yields are above 90% in the range of 0.5-0.8 M salt. The specific activity (alpha-1 activity/A280), an indication of purity, increases by 20-40% over the initial activity at pH 8 with no dependence on salt concentration (Figure 9). Adjustment of pH in future experiments may give increased purity.

The concentration of PEG in the salt phase is plotted in Figure 10. The lowest values are obtained at 1.2 M salt. Reducing the temperature to -5 C further reduces the PEG levels by a factor of 2-10, giving a minimum of 0.06 mg/ml. At 0.8 M salt, higher concentrations of PEG are measured in 25 ml. vessels as compared to 100 ml. It was observed that small beads of PEG adhered to the walls of the smaller vessels. Thus, the increased ratio of surface area / volume increased the apparent PEG concentration. It is expected that all PEG values would decrease as the vessel volume is increased. Other materials (proteins) were not subject to carryover.

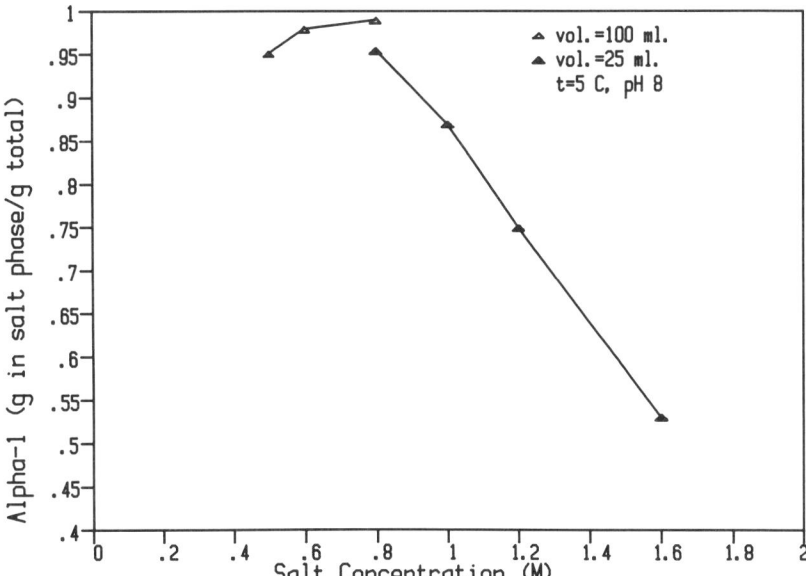

Figure 8. Recovery yield of Alpha-1 in the salt phase as a function of salt concentration.

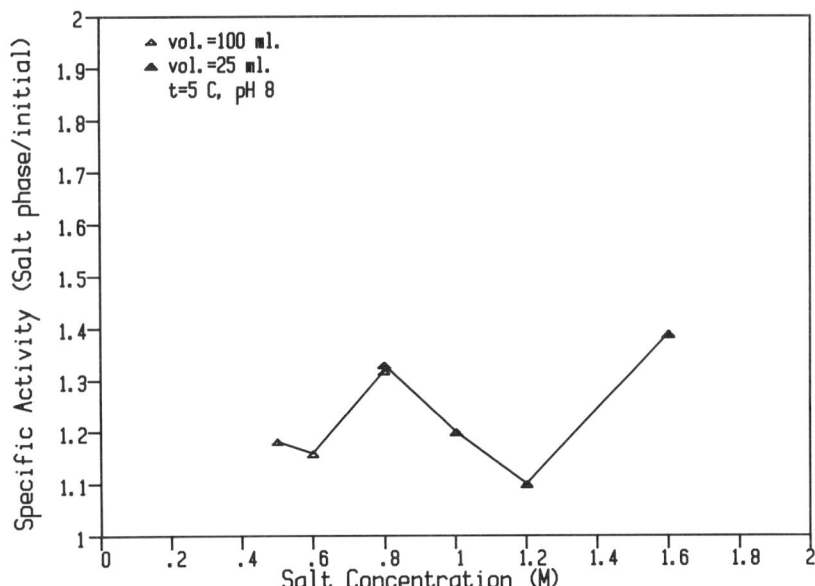

Figure 9. Change in specific activity of Alpha-1 (salt phase/initial) as a function of salt concentration.

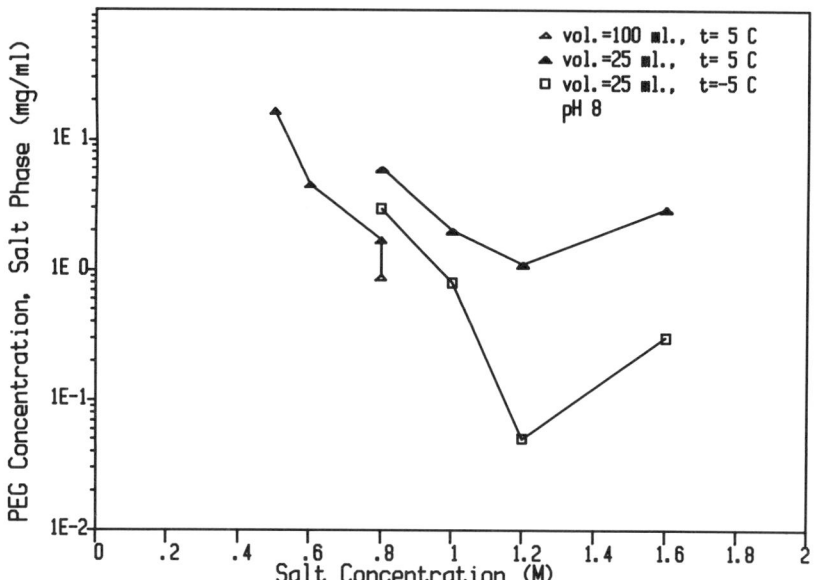

Figure 10. PEG concentration in the salt phase as a function of salt concentration.

The ability to distribute proteins and polymers, including nucleic acids, between two immisible phases has been shown. The liquid system consists of a PEG rich top phase and a salt rich bottom phase. Proteins of interest in industrial pharmaceuticals include nucleic acids (DNA), albumin, immunoglobulins, and alpha-1 antitrypsin. The initial purification of a single component (alpha-1) from a complex mixture (plasma) has been demonstrated. Further, residual PEG in this step has been severely reduced. An increasing number of new processes are utilizing PEG as an agent in precipitation or other operation. Removal of PEG from aqueous solutions by column chromatography, diafiltration, and other conventional methods is difficult. It has been shown that the concentration of PEG may be reduced by controlling physical parameters, even in complex mixtures.

The technique is applicable to an infinite variety of complex separations with unique attributes. The properties forming the basis of separation (surface charge, relative affinities) are unlike those of other commonly used processes and thereby complement other techniques to provide increased selectivity. High yield, low capital requirements and simple processing steps facilitate incorporation into a detailed purification scheme.

Legend of Symbols

K: partition coefficient = $C(t) / C(b)$.
C(t): Concentration in top phase, typically of a protein.
C(b): Concentration in bottom phase.
PEG: polyethlylene glycol.
a: alpha
b: beta
IgG: immunoglobulin G.
IgM: immunoglobulin M.
A-1, alpha-1: alpha-1 antitrypsin.
M: molecular weight (g / gmole).
x: factor in Bronsted formula (atmospheres x liters / g).
R: gas constant: (0.08205)(atmosphers x liters / gmoles x oK).
T: temperature (o K).

Literature Cited

1. Maniatis, T., Fritsch, E. F., Sambrook, J. "Molecular Cloning: A Laboratory Manual"; Cold Spring Harbor Laboratory; 1982; p. 458.
2. Albertsson, P.A. "Partition of Cell Particles and Macromolecules"; Wiley Interscience: New York; 1960.
3. Ryden, J., Albertsson, P. A. J. Coll. Interface Sci. 1971, 37, 219.
4. Hustedt, H., Kroner, K.H., Menge, U., and Kula, M-R., Trends in Biotechnology 1985, 3, no. 6, 139-141.
5. Backman, L., Shanbhag, V. P., Anal. Biochem. 1984, 138, 372.
6. Menge, U., Morr, M., Mayr, U., Kula, M. R., J. Appl. Biochem. 1983, 5, 75.
7. Albertsson, P. A., Biochemistry 1973, 12, 2525.
8. Johansson, G., Biochim. Biophys. Acta 1970, 222, 381.
9. Chaabouni, A., Dellacherie, E., J. Chrom. 1979, 171, 135.
10. Flanagan, S. D., Barondes, S. H., J. Biol. Chem. 1975, 250, 1484.

11. Mattiasson, B., Ling, T. G. I., *J. Immunol. Methods* 1980, 38, 217.
12. Shanbhag, V. P., *Biochim. Biophys. Acta* 1973, 320, 517.
13. Busby, T. F., Ingham, K. C., *Vox Sang.* 1980, 39, 93.
14. Bungay, H. R., "Enzyme Engineering"; E. K. Pye and H. H. Weethall, ed.; Plenum: New York; 1978, vol. 3, p. 225.
15. Hofsten, B. v., Baird, G. D., *Biotech. Bioeng.* 1962, 4, 403.
16. Albertsson, P. A., *Biochim. Biophys. Acta* 1965, 103, 1.
17. Alberts, B. M., "Methods in Enzymology, Nucleic Acids"; Wiley Interscience: New York; 1967, vol. 12, 566.
18. Kroner, K. H., Hustedt, H., Granda, S., Kula, M. R., *Biotech. Bioeng.* 1978, 20, 1967.
19. Kula, M. R., Kroner, K. H., Hustedt, H., and Schutte, H., *Ann. NY Acad. Sci.* 1981, 369, 341.
20. Schutte, H., Kroner, K. H., Hummel, W., Kula, M. R., *Ann. NY Acad. Sci.* 1983, 413, 270.
21. Hustedt, H., Kroner, K. H., Stach, W., Kula, M. R., *Biotech. Bioeng.* 1978, 20, 1989.
22. Andersson, E., Johansson, A-C., Hahn-Hagerdal, B., *Enzyme and Microb. Tech.* 1985, 7, July, 333-338.
23. Veide, A., Smeds, A. L., Enfors, S. O., *Biotech. Bioeng.* 1983, 25, 1789.
24. Kroner, K. H., Hustedt, H., Kula, M. R., *Proc. Biochem.* 1984, 19, 170.
25. Craig, L. C., Craig, D., "Techniques of Organic Chemistry"; ed. A. Weissberger; Interscience Publishers: New York; 1956, vol.3.
26. Higgins, J. J., Lewis, D. J., Daly, W. H., Mosqueira, F. G., Dunnill, P., Lilly, M. D., *Biotech. Bioeng.* 1978, 20, 159.
27. Bradford, M., *Analytical Biochemistry* 1976, 72, 248-254.
28. Coan, M. H., Brockway, W. J., Eguizabel, H., Krieg, T., Fournel, M., *Vox Sang.* 1985, 48, 333.
29. Burton, K., *Biochem. J.* 1956, 62, 315.
-30. Giles, K. W. and Myers, A., *Nature, London* 1965, 206, 93.
31. Dove, G. and Naab, P., personal communications.
32. Albertsson, P.A., *Advances in Protein Chemistry* 1970, 24, 309.
33. Bronsted, J. N., *Z. phys. Chem., A.* (Bodenstein-Festband) 1931, 257; from ref. 2.
34. Bronsted, J. N., Warming, E., *Z. phys. Chem., A.* 1931, 155, 343; from ref. 2.

RECEIVED March 26, 1986

Modeling of Precipitation Phenomena in Protein Recovery

C. E. Glatz and R. R. Fisher

Department of Chemical Engineering, Iowa State University, Ames, IA 50011

> A review of efforts to experimentally characterize and model the phenomena important in protein precipitation shows that, despite successes, continued work is necessary to produce accurate mechanistic descriptions of this method of protein recovery.
> The formation and growth of the primary particle in acid precipitation has been described in terms of the protein supersaturation. Aggregate growth by collision results in a size-dependent rate expression. Aggregate breakage, by shear or collision, remains to be adequately described in light of recent work. Population balances serve to model the combined phenomena.
> Recent work identifies mixing during precipitant addition as a determinant of aggregate physical properties; such effects are described with a floc-strength model.

Reaping the benefits of the new biology and even the continued development of traditional biotechnology poses problems in several areas. Two of these, synthesis of the desired product and its end use, have been and will continue to be the focus of much research. Relatively neglected has been the recovery and purification of these biological products, the intermediate steps that constitute the area of "downstream processing." It is this last area that is proving to require the greatest effort in practice and that has the poorest base of fundamental engineering understanding on which to draw.

The topic of this paper is the modeling of events occurring in the recovery of proteins and in the conditioning of the product streams for further purification using precipitation. The typical goal of downstream processing is the recovery of a desired product from a very dilute stream while minimizing the loss of the material in what is usually a multi-step separation process. Precipitation enables an early concentration of the product and can simultaneously serve to remove contaminants that would interfere with subsequent purification steps. Further, the wide variety of potential precipitating agents that exists permits selection of the particular agent capable of recovering the target species under conditions where activity is retained.

The target species considered here are proteins, and the principles developed may be applied to any protein-containing aqueous stream, including fermentation broths, plant extracts, and waste streams, whether the material is destined for food, pharmaceutical, or chemical application.

Protein Precipitation

The stability of proteins in solution is determined by a number of factors that govern protein-protein, protein-solvent, and solvent-solvent interactions. Sufficient alteration of any of these interactions can result in dramatic reductions in solubility. Hence, proteins can be precipitated by a variety of agents including organic solvents, divalent cations, heat, acids/bases (pH adjustment), salts, nonionic polymers (eg. polyethylene glycol), and polyelectrolytes. These means of altering solubility have been known and used for some time. What had been lacking was a description of the mechanism of formation of the particulate phase, the environmental determinants of the characteristics of this phase, and the connection between these characteristics, particulate behavior in the subsequent purification steps, and retention of functional activity. There was, therefore, little knowledge on which to base design of the precipitation stage so that the precipitate would be easily recovered, the maximum amount of protein would be in its native or active state, and as many contaminants as possible would be removed.

Recent research, the bulk of which has been gathered from study of the isoelectric precipitation of soy protein, has provided a good deal of this missing information.

Grabenbauer and Glatz (1) and Virkar et al. (2) have shown that precipitation proceeds by an initial rapid formation of submicron primary particles followed by collision-controlled aggregation of these primary particles. The latter growth phase is complicated by the simultaneous shear-controlled breakup of the aggregates. We will examine each step in turn, including modeling approaches for each.

Primary Particle Formation. The initial stage of primary particle formation had been observed by Parker and Dalgleish (3) for enzymatically destabilized casein. They used light scattering and turbidity measurements to follow the weight-average molecular weight (M_w) of the associating casein particles. After an initial period of accelerating rate, the kinetic behavior could be described by von Smoluchowski's theory of perikinetic coagulation. The result in terms of M_w was

$$M_w = M_o + 2 \text{ wkt} \tag{1}$$

where M_o is the molecule weight of the individual proteins, k is the coagulation rate constant, w is the concentration (weight basis) of protein, and t is time.

Nelson and Glatz (4) examined the role of environmental conditions in determining the size and number of these primary particles. They found size to depend on the precipitating agent

(HCl, H_2SO_4, or Ca^{2+}) and total protein concentration, but not on mixing conditions. Their conclusion was that primary particle growth is governed by supersaturation-controlled nucleation/growth phenomena.

This mechanism for the formation of primary particles can be described using Nielsen's (5) expressions for homogeneous nucleation with diffusion-controlled growth in precipitation. In his discussion, the nucleation rate, $J(c)$, is expressed as a power-law function of supersaturation, c,

$$J(c) = k_n c^m \tag{2}$$

where k_n is the nucleation rate constant and m is the nucleation power constant. The number of primary particles per unit volume, N_1, formed in a batch precipitation can be calculated as

$$N_1 = \int_0^\infty J(c)\,dt = k_n \int_0^\infty c^m\,dt \tag{3}$$

Supersaturation, c, may be expressed as a function of initial supersaturation, c_o,

$$c = (1-\alpha)\,c_o \tag{4}$$

where α is the fraction of supersaturated protein that has been precipitated. Combining a diffusion-controlled growth-rate expression with the assumption of spherical primary particles gives the volumetric growth rate of formed particles as first order in supersaturation

$$\frac{dV}{dt} = 4\pi D_m r c v \tag{5}$$

where V is the particle volume, D_m is the protein diffusivity, v is the protein molecular volume, and r is the molecular radius. Solving Equations 3-5 together with an overall mass balance, Nielsen obtained the approximate result

$$N_1 = k_n^{3/5}\,D_m^{-3/5}\,\left(\frac{4\pi}{3}\right)^{-2/5}\,v^{-1/5}\,c_0^{(3m-1)/5}\,C_{m,D} \tag{6}$$

where $C_{m,D}$ is a weak function of m, approximately equal to one. Hence the stronger the dependence of nucleation on supersaturation, the greater will be the increase in number of primary particles as initial supersaturation increases. For $(3m-1)/5 > 1$ (i.e. $m > 2$), the size of those particles will decrease with initial supersaturation. No dependence on mixing conditions appears; the concentration dependence for soy protein precipitates (via hydrochloric acid addition) was found (4) to be

$$N_1 = 2.67 \times 10^{11}\,c_o^{0.84} \tag{7}$$

requiring $m = 1.7$. Over the range of concentrations studied (0.15 -

30 kg/m^3) the size of primary particles increased from 0.16 to 0.27 μm.

Aggregate growth and breakup. The primary particles (produced as described above) form the starting material for the aggregation stage. This stage parallels what occurs in other aggregation/coagulation/flocculation systems. It differs from many of these however, in that the aggregates are particularly prone to breakup and their size is smaller than the Kolmogorov microscale of turbulence, subjecting them to different controlling fluid forces during growth and breakup (6). Since particle size is one of the determinants of the efficiency of solid-liquid separations (filtration rate and settling velocity are both proportional to the square of particle diameter (7)), the modeling and characterization of the particle size distributions is important.

Population balances were combined with the proposed mechanisms to model the size distributions from continuous stirred precipitators. The postulated failure of collisions between larger aggregates to form lasting agglomerates reduces the growth process to one where only primary particles and small aggregates can serve as growth units, though larger sizes may serve as collectors. Modelled in this way, growth becomes a continuous process. Particle size distributions have been successfully modelled over a wide range of conditions for continuous stirred-tank precipitators (1, 6, 8).

Models of the particle size distribution

Asssumptions. The mathematical models based on the population balance incorporate the following physical features and simplifying assumptions:

1. Protein comes out of solution very quickly and therefore an accounting is needed only for the solid material. This is supported by tubular reactor studies (2, 9) where precipitation, in terms of removal of soluble protein, is completed within 1 s for the protein concentrations above 2 kg/m^3.

2. Growth of an aggregate occurs by collision with primary particles and smaller aggregates. However, collisions between larger aggregates are ineffective in forming lasting aggregates. Gregory (10) has shown that collision efficiency decreases considerably with increasing size of equal-sized colliding species. In addition, aggregate-aggregate attachments would be relatively weak as the result of the lower bond densities at these points compared to bond densities within aggregates. A ratio of 10 to 1 has been reported as typical (11) for the ratio of contacts within an established aggregate to contacts between two such aggregates. Growth is therefore viewed as the incremental addition of small units to the growing aggregates.

3. The effectiveness of these collisions of small particles with growing aggregates is independent of the size of the growing species.

Population Balances. Three different models based on two approximations regarding the mode of breakage and two approximations regarding the size dependence of growth rate have been examined. The differential equations for modeling the size distribution are based on a population balance on aggregates of size L which, for a CSTR at steady state, mean residence time τ, and with no particles in the feed, reduces to

$$\frac{d(Gn)}{dt} + \frac{n}{\tau} + D - B = 0 \qquad (8)$$

where n is the number density of particles, τ is the reactor mean residence time, and where G, D, and B are the rates for differential growth, volumetric death by breakup, and volumetric birth by breakage of larger particles, respectively; n, G, D, and B may be functions of L.

The first model incorporates expressions for each of these terms (6), in which growth is approximated as linear with aggregate size;

$$G = A\phi_1 V_g L = K_0 L \qquad (9)$$

where A is a constant incorporating collision effectiveness, ϕ_1 is the volume fraction of primary particles, V_g is the root-mean-square velocity gradient, and K_0 is the product of these three, called the growth rate constant.

Breakage is described by

$$D = k'V_g \left(\frac{\mu V_g}{\sigma_{ya}}\right)^\delta n = kL^\beta n \qquad (10)$$

where k' and k are death-rate constants, μ is the solution viscosity, σ_{ya} is the aggregate yield stress, and δ and β are breakage power constants.

Breakup requires terms accounting for the sudden disappearance (death) of parent aggregates and corresponding appearance (birth) of daughter fragments. The first and second models assume that aggregates break up to form a small number of daughter fragments of significant mass. The number of daughter fragments would tend to be greater for larger parent aggregates; this is approximated as an average fragment number, f, dependent on the mean size of the distribution. Daughter fragments from a given parent are assumed to be of equal volume. This gives

$$B = fD(f^{1/3}L) \qquad (11)$$

The resulting equation relating number density to size is

$$\frac{dn}{dL} = \frac{k}{K_0} L^{\beta-1} \left(f^{(\beta/3)+1} n(f^{1/3}L) - n\right) - \frac{n}{L}(1 + \frac{1}{\tau K_0}) \qquad (12)$$

Summarizing, the model parameters are K_0, k, β, and f. The death and birth expressions assume breakage into equal-sized fragments such that total particle volume is conserved and that a power law describes the increased susceptibility of aggregates to breakup as size increases. The growth term assumes that collision due to the spatial variation of turbulence is the predominant factor, as has been demonstated (e.g. (<u>12</u>)).

The second model considered here (<u>1</u>) results from assuming that the growth rate, G_m, is independent of size, giving

$$\frac{dn}{dL} = \frac{K}{G_m} L^\beta \left(f^{(\beta/3)+1} n(f^{1/3}L) - n \right) - \frac{n}{G_m \tau} \tag{13}$$

The boundary condition used for Equations 12 and 13 is that the calculated total volume of aggregates equals the measured aggregate volume.

<u>Parameters.</u> Past work has suggested that for a given protein concentration, β and f could be fixed at reasonable values, leaving only two fully variable parameters. The method of solution is discussed elsewhere (<u>8</u>). Figure 1, from data of Brown and Glatz (<u>15</u>), shows that breakup of large aggregates results in two or more main fragments and a number of small fragments comprising relatively little mass. The latter can be neglected in the balance, although they will serve to increase ϕ_1 as they are considered capable of being growth units. The larger aggregates are expected to form the greater number of daughter fragments. However, Pandya and Spielman (<u>13</u>) found that allowing for this within a given distribution of kaolin-Fe(OH)$_3$ flocs was no better than using a constant $f = 2.5$.

The model based on a growth rate independent of aggregate size, Equation 13, gave reasonable fits by residual sum of squares criteria except at high protein concentrations. However, at all protein concentrations studied, the predicted curves were biased in the manner in which they deviated from experimental observation. This was particularly evident at small sizes where the local minimum/maximum traits were largely lost. The first model, based on a growth rate linear in aggregate size, Equation 12, gave satisfactory fits of the particle size data (in fact, for most runs the model fit the experimental points more closely than did the six-parameter Chebyshev polynomial on which the model fitting was actually based) as well as successfully describing the local minimum/maximum traits. The curves presented in Figure 2 are based on this model, using the data of Glatz et. al. (<u>8</u>) who discuss the behavior of model parameters k and K_0 at different reactor conditions.

The third and final particle size distribution model assumes that growth is linear as in the first but that breakup results in predominantly small particles (thorough breakage) which are too small to measure by the electronic particle counters used to characterize the suspension. Petenate and Glatz (<u>6</u>) have provided analytical solutions for this model.

The focus of the above modeling has been on continuous stirred-tank reactors. The general principles have been extended to interpret results from batch and tubular reactors, as well, though detailed modeling has not yet been attempted (<u>8</u>).

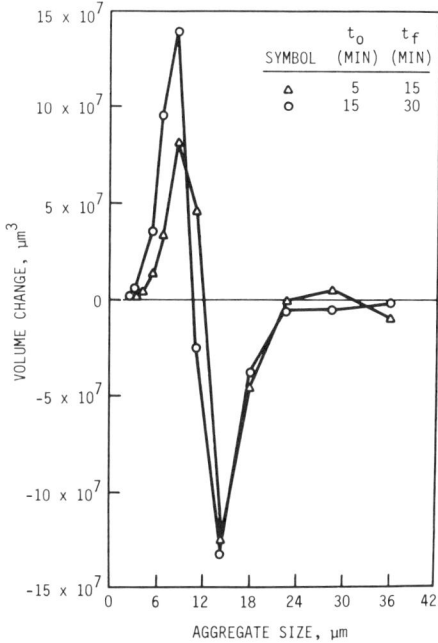

Figure 1. Plot of change in aggregate volume vs. aggregate size for given time interval during breakup of isoelectrically precipitated soy protein. Particle volume fraction, 0.00531. Shear rate, 1010 s^{-1}.

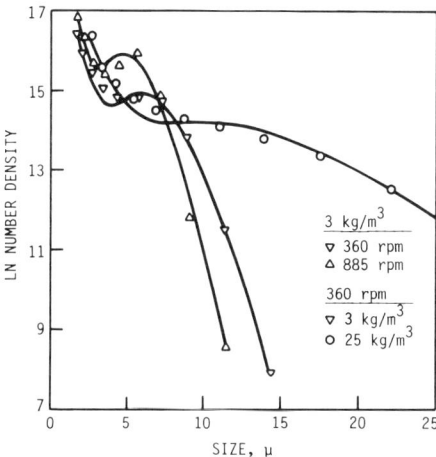

Figure 2. Particle (number) size distributions for isoelectrically precipitated soy protein showing the effects of shear rate and protein concentration. Points are experimental data; curves are the model fit using Equation 12. Shear rates: $\triangle$, 417 s^{-1}; $\triangledown$, 108 s^{-1}; o, 85 s^{-1}.

Previous attempts have been made to model size distributions allowing for growth by collisions of all aggregate-aggregate combinations. Such models (2, 14) predicted much higher growth rates than were observed, offering further evidence for the ineffectiveness of aggregate-aggregate collisions.

The modeling equations, when all particle-particle collisions are assumed effective, are represented (in a batch reactor) by

$$\frac{dN_k}{dt} = \frac{1}{2}\left(\sum_{\substack{i=1\\j=k-i}}^{i=k-1} 4/3(r_i+r_j)^3 \, V_g N_i N_j\right)$$

$$- N_k\left(\sum_{i=1}^{\infty} 4/3(r_i+r_k)^3 \, V_g N_i\right) \tag{14}$$

where N, the number concentration of particles, and r, the particle radii, are specified for particle sizes i, j, and k such that $r_k^3 = r_i^3 + r_j^3$.

This equation replaces Equation 8, but does not explicitly account for aggregate breakage. Virkar (2) introduced breakage to this balance by not allowing collisions that would result in a particle larger than a fixed maximum size.

Breakage Models. We are continuing our study of stirred tank behavior of isoelectric precipitates by examining the breakup phenomena and the modeling equation for breakup in more detail. Data are interpreted in the light of three proposed treatments of breakup (15). Two are based on breakup under fluid shear, using the concepts of a maximum stable size (16-18) and similarity (19-20). The third is based on collisional breakage which has been discussed but not observed by Glasgow and Luecke (21), and thought to occur with protein aggregates in laminar shear (22).

Other Precipitants. Extension and modification of these modeling efforts will be required for their application to precipitations other than isoelectric. Other low molecular weight precipitants have been shown to result in the same sort of aggregate morphology (Ca^{2+}; (4)) and growth kinetics (ethanol, Ca^{2+}, $(NH_4)_2SO_4$; (9)). For these precipitants no modification should be necessary, although the dependence of aggregate strength on the physicochemical conditions will change and with it the strength-dependent model parameters. Based on very preliminary results (23) the behavior of polyethylene glycol is also expected to be quantitatively similar.

Precipitate Behavior

Beyond describing what is occurring in the precipitation step itself, work has been done in relating the characteristics of the material leaving the precipitator to its behavior in subsequent operations. The importance of aggregate "aging" to condition the aggregates to resist breakup during shear encountered in pumps and centrifuges is documented (24-25). The influence of preparation conditions on centrifuge capacity and sludge rheology has been observed (26), an explanation for aggregate strength and rheological properties proposed (4) in terms of an elastic floc model (11), and the preparation- (27) and aging- (14) dependent settling behavior reported.

The work of Fisher et al. (27) illustrates some design considerations beyond that of the particle size at the precipitation outlet. These workers were concerned with the mixing conditions during the addition of acid to the protein solution. Local extremes in pH are known to cause irreversible denaturation of proteins (28) which will alter their precipitation behavior. Further, Hoare (29) has reported that the precipitate properties differ with extremes in operating conditions. One extreme exploited in inorganic precipitations is homogeneous precipitation (30), which involves the homogeneous production of the precipitant, usually by a controlled chemical reaction, within the solution. Advantages of this include the production of denser precipitate and reduced coprecipitation. Fisher et al. (27) approximated homogeneous isoelectric precipitation by the addition of acid from a dialysis bag suspended in the protein solution on a rotating shaft. The degree of fractionation of two proteins (glycinin, pI = 6.0, and β-conglycinin, pI = 4.8) from a total soy extract and the physical characteristics of the precipitates were contrasted with the same properties of precipates formed during rapid acid addition. Precipitate fractions were taken at pH 6.0 and 4.8.

The compositions of the fractions, Table I, indicate that separation of the glycinin and β-conglycinin did occur, with substantial enrichment of the glycinin phase in the pH 6.0 fraction. Table I also shows that no differences in the fractionation of glycinin or β-conglycinin can be attributed to the mixing during acid addition.

In hindered settling tests on the pH 4.8 product the aggregate prepared by slow acid addition clearly settled faster than that prepared by rapid acid addition. Since aggregate size and density, as measured prior to settling, could not account for the different settling rates, another explanation was sought. It was concluded that the controlling characteristic of the hindered settling is the aggregate's ability to aggregate further and become large enough to settle.

The characteristics of the isoelectrically precipitated aggregates were interpreted in light of a model of floc stength as developed by Firth and Hunter (11) and applied by Nelson and Glatz (4). This model holds that the strength, σ_{ya}, of an aggregate of primary particles is a product of the number of bonds per area and the attractive force per bond. They further showed that

$$\sigma_{ya} = \frac{Q\phi_1}{d_1} \tag{15}$$

where Q is an interaction potential function--the sum of charge-charge repulsive and van der Waals attractive contributions--and where d_1 is the diameter of the primary particle. The application of this model helped to identify orientation of the protein during its incorporation into the primary particle as a determining step in the subsequent strength of the aggregate.

Table I. Physical and Compositional Properties of the Protein Extract and the Rapid and Slow Protein Precipitates

Property	Total protein extract[a]	pH 6.0 Rapid	pH 6.0 Slow	pH 4.8 Rapid	pH 4.8 Slow
% glycinin in fraction[b]	27.6	49.3	52.6	10.0	8.1
%β-conglycinin in fraction[b]	16.7	4.0	4.1	27.4	21.8
d_1[c] (μm)	--	0.13	0.19	0.51	0.36
d_{50}[d] (μm)	--	--[e]	--	11.31	7.88
hindered settling time (min)[f]	--	--[g]	--	17	7

[a] Determined by biuret colorimetric assay.

[b] Immunologically active protein expressed as percent of total protein, determined by rocket-gel immunoelectrophoresis.

[c] Primary particle diameter, measured from scanning electron micrographs of the aggregate.

[d] Mean aggregate diameter (on a volume basis), determined from particle size distribution.

[e] The pH 6.0 aggregates were too weak to be characterized from size distribution.

[f] Time for interface to drop to 30% of initial slurry height.

[g] The pH 6.0 aggregates did not settle in the hindered settling test.

Conclusion

Models for continuous, stirred precipitation behavior are at a reasonably successful stage, though the expression for breakup can be improved in the light of recent studies. Batch and tubular reactor models must additionally include an explicit accounting for primary particle formation and behavior of small particles. Some of the data necessary to do so have been collected, but number density data at the small sizes is still lacking. Finally it remains to be seen how successful the models developed for isoelectric precipitation will be in describing precipitation with other classes of precipitants.

Acknowledgments

This work was supported by the Engineering Research Institute of Iowa State University through National Science Foundation Grant No. CPE-8120568.

Literature Cited

1. Grabenbauer, G. C.; Glatz, C. E. Chem. Eng. Comm. 1981, 12, 203-219.
2. Virkar, P. D.; Chan, M. Y. Y.; Hoare, M.; Dunnill, P. Biotechnol. Bioeng. 1982, 24, 871-887.
3. Parker, T. G.; Dalgleish, D. G. Biopolymers 1977, 16, 2533-2547.
4. Nelson, C. D.; Glatz, C. E. Biotechnol. Bioeng. 1985, 27, 1434-1444.
5. Nielson, A. E. "Kinetics of Precipitation"; Macmillan Co.: New York, 1964, Chap. 7.
6. Petenate, A. M.; Glatz, C. E. Biotechnol. Bioeng. 1983, 25, 3059-3078.
7. McCabe, W. L.; Smith, J. C. "Unit Operations of Chemical Engineering, 3rd Edition"; McGraw-Hill, Inc.:New York, 1976; p. 938-971.
8. Glatz, C. E.; Hoare, M.; Landa-Vertiz, J. AIChE J. 1986, in press.
9. Chan, M. Y. Y.; Hoare, M.; Dunnill, P. Biotechnol. Bioeng., 1985, accepted for publication.
10. Gregory, J. Chem. Eng. Sci. 1981, 36, 1789-1794.
11. Firth, B. A.; Hunter, R. J. J. Colloid Interface Sci., 1976, 57, 266-275.
12. Yuu, S. AIChE J. 1984, 30, 802-807.
13. Pandya, J. D.; Spielman, L. A. J. Colloid Interface Sci. 1982, 90, 517-532.
14. Hoare, M. Trans. Inst. Chem. Eng. 1982, 60, 157-163.
15. Brown, D. L.; Glatz, C. E. Paper presented at the AIChE Annual Meeting 1985, Chicago.
16. Tomi, D. T.; Bagster, D. F. Trans. Inst. Chem. Eng. 1978, 56, 1-8.
17. Parker, D. S.; Kaufman, W. J.; Jenkins, D. J. J. Sanit Eng. Div. ASCE 1982, 98, 79-99.
18. Tambo, N.; Hozumi, H. Water Res. 1979, 13, 421-427.
19. Ramkrishna, D. Chem. Eng. Sci. 1974, 29, 987-992.

20. Narsimhan, G.; Ramkrishna, D.; Gupta, J. P. AIChE J. 1980, 26, 991-1000.
21. Glasgow, L. A.; Luecke, R. H. Ind. Eng. Chem. Fundam. 1980, 19, 148-156.
22. Twineham, M.; Hoare, M.; Bell, D. J. Chem. Eng. Sci. 1984, 39, 509-513.
23. Glatz, C. E., unpublished data.
24. Bell, D. J., Dunnill, P. Biotechnol. Bioeng. 1982, 24, 1271-1285.
25. Hoare, M.; Narendranathan, T. J.; Flint, J. R.; Heywood-Waddington, D.; Bell, D. J.; Dunnill, P. Ind. Eng. Chem. Fundam. 1982, 21, 402-406.
26. Bell, D. J.; Dunnill, P. Biotechnol. Bioeng. 1982, 24, 2319-2336.
27. Fisher, R. F.; Glatz, C. E.; Murphy, P. A. Biotechnol. Bioeng. 1985, accepted for publication.
28. Salt, D. J.; Leslie, R. B.; Lillford, P. J.; Dunnill, P. Eur. J. Appl. Microbiol. Biotechnol. 1982, 14, 144-148.
29. Hoare, M. Trans. Inst. Chem. Eng. 1982, 60, 79-87.
30. Berg, E. W. "Physical and Chemical Methods of Separation"; McGraw-Hill, Inc.:New York, 1963; pp. 279-284.

RECEIVED March 26, 1986

PURIFICATION OPERATIONS

10

Process Considerations for Scale-up of Liquid Chromatography and Electrophoresis

S. R. Rudge and M. R. Ladisch

Laboratory of Renewable Resources Engineering, School of Chemical Engineering, and Department of Agricultural Engineering, Purdue University, West Lafayette, IN 47907

> Chromatography is an important preparative and industrial process. Scale-up of chromatographic processes requires computation of mass transfer characteristics as a function of column area, support particle size and feed volume. In this context, an analytical solution for longitudinal diffusion in packed beds, developed by Lapidus and Amundson, is used to demonstrate the characteristics of a typical size exclusion separation of proteins, including estimation of maximum sample size as allowed by support properties. Electrophoresis is also a powerful fractionation technique for proteins, but is subject to many microscopic effects. These include electric double layers, hydrodynamic drag, and electrical relaxation. In addition, macroscopic effects, such as electroosmosis and thermal gradients, also impact separation efficiency. These effects are discussed in relation to elution processes using selected examples. The combination of an electric field with a chromatographic process has recently been proposed to extend the power of electrophoresis separations. Analysis of such a process, referred to as electrochromatography, is also presented.

Mixtures can be separated by taking advantage of a physical property which varies among the mixture's components. Such properties include boiling points, equilibria with other substances, solubilities, isoelectric points, size, and density. These physical properties can be exploited by manipulating thermodynamic variables, which include temperature, pressure, ionic strength, pH, and electric potential. While all of these variables have been used to separate compounds, there has recently been a great deal of interest in the effect of electric potential on separations, particularly in biological systems, as well as application of process chromatography to a variety of molecules. This chapter discusses both the key factors which impact the practical implementation of chromatography, and the current thinking on its combination with electric potential to improve resolution.

Many industrial or preparative separation processes are elution processes. In elution processes a pulse of sample followed by eluent is continuously fed, and another stream is continuously recovered. However, product leaves the system in pulses, making this process a batch process. This is different from a continuous process, in which feed and product streams are continuously fed and recovered, respectively, and are at steady state. Semi-continuous processes alternate between continuous product and regeneration cycles. Simple fixed bed adsorption is an example of a semi-continuous process, while distillation is an example of a continuous process. Chromatography is typically an elution process although continuous forms of this separation have been reported (1-4).

Elution processes allow introduction of feed and withdrawal of product without irreversible changes in stationary media. Elution of products is usually driven by the flow of a fluid, although other driving forces are conceivable. Fluid driven elution is readily controlled on a process scale, especially in the case of incompressible fluids. Pressure drop is always a consideration when a fluid is driven, and becomes more important with increasing equipment length.

The Capacity Factor in Chromatography

One challenge in process chromatography applied to biologically active or derived molecules is that these molecules exist with many other co-products, particularly if the desired product molecule is fermentation derived (5). In a separation scheme, a target molecule can be separated from other components based on differences in shape or size, ionic character, hydrophobic character, and/or bio-affinity (6). The resin's capacity to interact with the molecule is a key parameter, and is described by the capacity factor:

$$k'_i = \frac{n'_i}{c'_i} = K'_i \frac{V_s}{V_m} \qquad [1]$$

where n'_i is the moles of solute i in the stationary phase, c'_i is the moles of solute i in the mobile phase; K'_i is the distribution coefficient; V_s is the fluid volume displaced by the stationary phase and V_m is the fluid volume of the mobile phase. Since $V_T = V_m + V_s$, we can define $\bar{\alpha} = \frac{V_m}{V_T}$ and V_s as $(1 - \bar{\alpha}) V_T$. Then

$$k'_i = K'_i \frac{1 - \bar{\alpha}}{\bar{\alpha}} \qquad [2]$$

If a pore fraction, ϵ, can be determined for a particular support, then these equations may be further refined:

$$k'_i = K'_i \frac{1-\bar{\alpha}}{\bar{\alpha}} = K_i \frac{\epsilon(1-\bar{\alpha})}{\bar{\alpha}} \quad [3]$$

K_i is described differently from K'_i and reflects only the volume available to fluid within the support. $K'_i = K_i \epsilon$, where $\epsilon = \frac{V_{pores}}{V_s}$. The difference between capacity factors of the desired solute and other products reflects the extent of separation which can be expected. When the capacity factor is large ($k'_i \gg 1$), the support takes on characteristics of an adsorbent. In this case, a solute will adsorb onto the support at the conditions of sample loading, and desorb only when conditions in the eluent, such as salt concentration, pH, polarity, or temperature, are changed. This fits the category for adsorption/desorption for which quantitative descriptions are available (7-10). This chapter primarily addresses cases where k'_i is small, i.e., chromatographic separations.

Bio-Affinity. An affinity support is an example of a support with a high capacity factor. This support has a spacer arm attached to a ligand with a highly specific affinity for a solute (11). The ligand acts here as an adsorbent in which a change of conditions in the bulk fluid surrounding the support is required to wash off the adsorbed solute. While quite elegant in theory, many times the product will be present in small concentrations, and nonbinding impurities will be present at much higher concentrations. If the ratio of product to other solutes is low (for example, 1 to 1000), even slight adsorption of nonbinding solute onto the matrix or ligand can translate into significant loss in selectivity. Effective use of an affinity support may therefore require sample clean up steps upstream as well as the appropriate selection of a matrix which is inert with respect to the solutes. Given current technology and the potentially high cost of affinity supports, such an approach suggests bio-affinity might be best considered in the context of a product polishing step (6).

Ion Exchange Resins As Chromatographic Supports. A versatile type of support is one which can exhibit differential ion exclusion, size exclusion, ion complexing, and hydrophobic interaction characteristics with respect to a variety of molecules. In fact, a generic support of this type exists, and is based on a copolymer of sulfonated styrene-divinylbenzene (DVB). Sulfonation impacts the ability of the support to complex counter-ions which, in turn, can complex with the solute or exclude an ionic species. The degree of resin cross-linking is proportional to DVB content and determines effective porosity of the resin and size exclusion characteristics.

The separation shown in Figure 1 illustrates the versatility of such a resin (12,13) packed in a 6 mm i.d. by 60 cm long jacketed column, maintained at 80^0C. In this case, separation of oligosaccharides G_7 to G_2

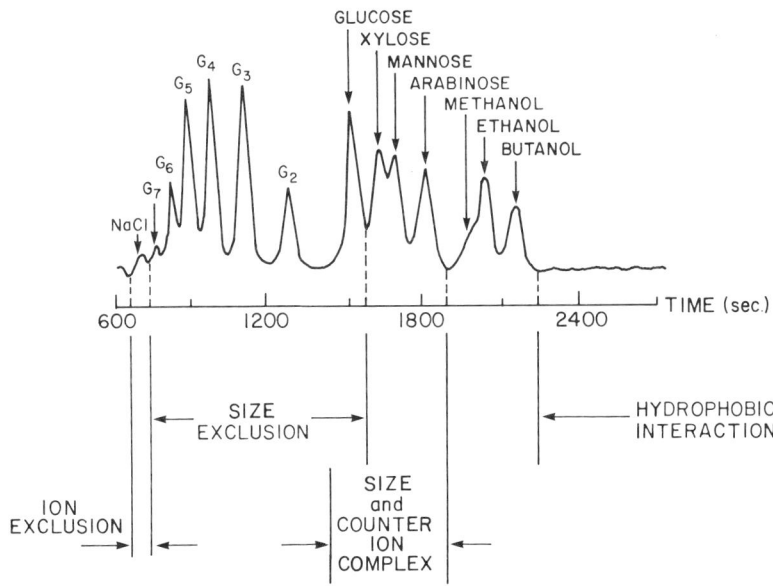

Figure 1. Ion exchange separation of oligosaccharides, sugars, and alcohols, conditions as given in text.

(molecular weight of 1152 to 342), monosaccharides (molecular weights of 180 and 150), and alcohols is shown. As indicated in Figure 1, ion exclusion (NaCl), size exclusion (G_7 through glucose, hexoses and pentoses), ion complexing (glucose/mannose and xylose/arabinose), and hydrophobic interactions (alcohols elute in order of increasing size and hydrophobic character) all contribute to the observed separation. The conditions for this separation were (12,13):

eluent: water
eluent velocity: 2.1 cm/min (based on cross-sectional area of empty column)
resin: Aminex 50WX4, Ca^{++}
sample size: 20 µl
detector: Waters differential refractometer
particle size: 20 to 30 microns

If scale-up can be achieved using larger particle size ranges (250 to 1000 micron), a significant potential for this type of support is indicated particularly when price ($100 to $500/cubic foot), availability of commercial quantities of cation ion exchange resins, and their history of use in the food and pharmaceutical industries are considered.

Scale-Up Considerations

Once a support having appropriate surface area, pore size, particle size, and surface characteristics is identified, engineering input for its use on a large scale is required. Analytical applications are characterized by a small sample size (less than 0.1% of column volume), low solute concentration (usually less than 1%), and use of small particle size (5 to 30 microns). In comparison, process-scale separations will probably be characterized by large sample size (up to 20% of column volume), high solute concentration (up to 30%); and relatively large particle size chromatographic supports (100 to 1000 micron). Thus, it may be more difficult to obtain clear resolution of the solutes on a process scale than it would be on an analytical scale. By careful control of process conditions, however, a simple column system (Figure 2) with a modest plate count (100 per meter based on glucose) can give satisfactory separation for relatively large sample volumes (Figure 3, reproduced from reference 14). Experience in our laboratory on a variety of column sizes ranging from 2 to 160 mm in diameter and 10 cm to 600 cm in length has shown that published semi-empirical correlations are useful in obtaining a first estimate of column performance (5,14,15).

Column Area. The desired sample volume to be applied to a column can be expressed in terms of column void volumes:

$$V_F = a\ V_m \qquad [4]$$

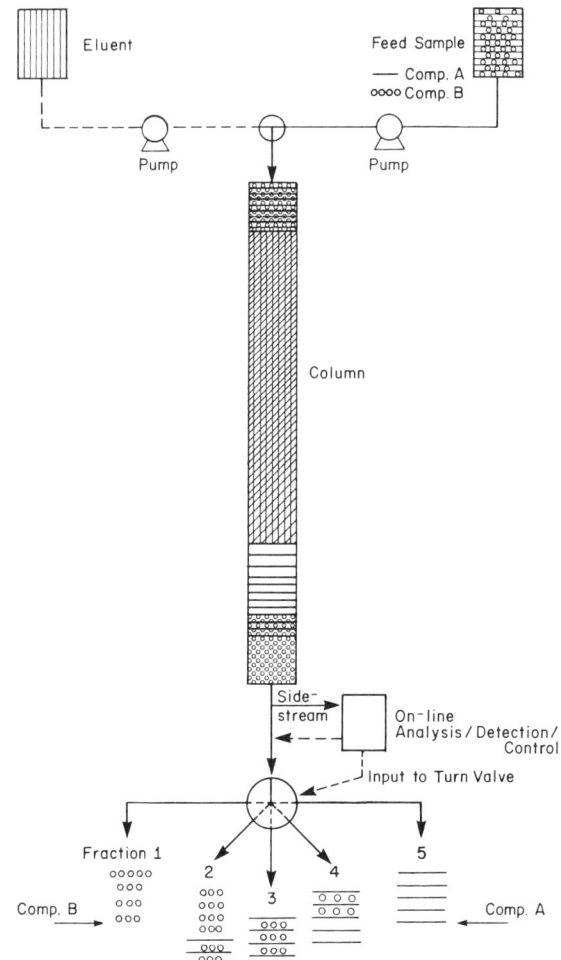

Figure 2. Chromatographic apparatus for process system.

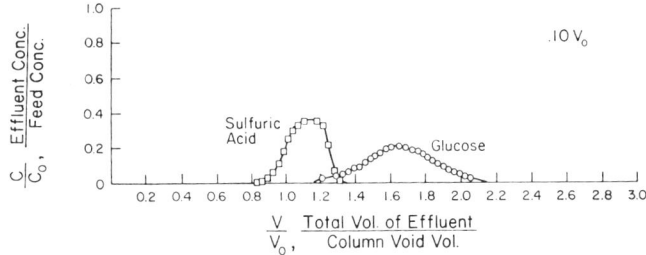

Figure 3. Ion exclusion separation of glucose and sulfuric acid. Sample size of 0.1 V_m, $T = 55$ C.

where V_F is sample volume; and a is dependent on support characteristics. For a column packed with support of a fixed particle size, operated at a fixed temperature and eluent linear velocity, the column cross-sectional area should increase in proportion to the sample size. However, proper column operation also depends on the ability to distribute the feed as a plug onto the column with as little back mixing as possible (5). As column diameter increases, this condition becomes more difficult to approximate due to practical problems in evenly distributing a liquid at velocities approaching those associated with creep flow. Hence, the length to diameter ratio, L/D, also becomes a factor. In general, an L/D of 10 or greater is desirable. An increase in length with an increase in column diameter may, however, be limited by the mechanical stability of the support and pressure drop. Special attention is then required for the liquid distribution systems and experimental trial and error becomes essential for scale-up.

Column Length. Column length causes an increase in dispersion of a solute due to both diffusion and backmixing caused by motion of the eluent. This results in an increase in the volume over which the sample will elute, relative to the volume of sample placed on the column. This phenomena is reflected by an increase in peak standard deviation (σ) where the width of a Gaussian peak is 4σ. In process chromatography, particularly in the case of sample overload, the peak may be skewed (16). A standard deviation can still be defined leading to a value of plate height, although moment analysis of the eluting peak must be used to obtain values of σ (5,17).

The concept of theoretical plates in chromatography is described elsewhere, and is not repeated here (18,19). Pieri et al. have described the application of the theoretical plate concept to column scale-up and demonstrated its utility in scale-up of pheromone separation over silica gel (15). We have found this same approach to be useful in separations over ion exchange resins used as chromatographic supports. In essence, this approach gives an estimate of column length for a given separation if the particle size is changed. This is based on empirical correlation of the number of theoretical plates, N_i:

$$N_i = \frac{D_i^{0.5}}{m} \frac{L}{v^{0.5} d_p^{1.5}} \qquad [5]$$

where D_i is a coefficient which reflects dispersion, L is column length, d_p is particle diameter, v is the eluent linear velocity, and m is an empirical constant. In this approach, the molecules separated are assumed to be chemically and physically similar. Therefore, D_i can also be assumed to be similar for these molecules as a first approximation (15). For the same linear velocity and plate count, column length will change with particle size as the 1.5 power:

$$L_x = L_A \left(\frac{d_{p,x}}{d_{p,A}} \right)^{1.5} \qquad [6]$$

Consequently, a 1 foot long column with a 50 micron support is equivalent to an 8 foot long column with a 200 micron particle size of the same support. For a binary system, resolution can be defined (19) by

$$R_s = \frac{(\phi-1)\, N^{1/2}}{4} \left(\frac{k'}{1+k'}\right) \qquad [7]$$

where

$$N = \frac{N_1 + N_2}{2} \qquad [8]$$

$$k' = \frac{k'_1 + k'_2}{2} \qquad [9]$$

$$\phi = \frac{k'_2}{k'_1} \quad \text{where } k'_2 > k'_1 \qquad [10]$$

Under process conditions, equations (5) to (10) are approximations when sample volumes and concentrations are larger than those usually associated with analytical chromatography. If support chemistry and physical characteristics are maintained through scale-up, and if the support is packed in a properly designed column, constant resolution is assumed to reflect a constant plate count. This is the basis for estimating column length by equation (6). Although approximate, this approach is useful for preliminary sizing in the absence of detailed data.

Theoretical Considerations In Size Exclusion Chromatography

In order to better understand chromatography it is important to study the underlying mass transfer operations which are occurring. These mass transfer phenomena are well studied (20-27), and analytical solutions exist to most limiting cases. In general, a mass balance for one component in a packed column is given by:

$$D\frac{\partial^2 c}{\partial x^2} = v\frac{\partial c}{\partial x} + \frac{\partial c}{\partial t} + \frac{1}{\alpha}\frac{\partial n}{\partial t} \qquad [11]$$

where

$$D\frac{\partial^2 c}{\partial x^2} = \text{dispersive flux } \left(\frac{mols}{time\ vol}\right)$$

Table I. Summary of Selected Initial and Boundary Conditions Applicable to Chromatography

Initial or Boundary Condition	Application	Ref.
1. Constant Inlet Concentration $c = c_a$ at $x = 0$, $t = 0$ $\dfrac{\partial c}{\partial t} = o$ at $x = 0$, $t > 0$ or $\dfrac{\partial c}{\partial t} = o$ at $x = 0$ $0 < t < t_{feed}$	When solutes are being eluted by an eluent of constant concentration, such as in linear chromatography, or when sample is being injected as a finite pulse.	21.25
2. No adsorption past column outlet $\dfrac{\partial c}{\partial x} = 0$ at $x = L$, $t > 0$	Used when D and v are very similar inside and outside the column and when accumulation is small compared to dispersion and convection.	20,22, 24
3. Initial concentration in column is 0 $c = 0$ at $t = 0$, $x > 0$ $n = 0$ at $t = 0$, $x > 0$	Used in nearly all cases. Applicable to all boundary conditions listed.	
4. Danckwert's Condition $D \dfrac{\partial c}{\partial x} - vc = vc_a$ at $x = L$, $t > 0$	Allows for backmixing at the column outlet by adding a diffusive flux term. Lets material diffuse into or out of the column if a gradient exists.	22-24, 26

$v \frac{\partial c}{\partial x}$ = convective flux ($\frac{mols}{time\ vol}$)

$\frac{\partial c}{\partial t}$ = accumulation of solute in mobile phase ($\frac{mols}{time\ vol}$)

$\frac{1}{\alpha} \frac{\partial n}{\partial t}$ = accumulation of solute on stationary phase ($\frac{mols}{time\ vol}$)

There are a variety of boundary and initial conditions which are useful for different situations. Some examples of these boundary conditions and their applications are given in Table I. A relationship between n (adsorbed species) and c (mobile species) must be found. These relationships may either be equilibrium or kinetic relationships (mass transfer rates). Some examples of equilibrium and mass transfer relationships may be found in Tables II and III, respectively. As pointed out by Lapidus and Amundson (25), equilibrium relationships in themselves are useful in cases where mass transfer rates are not limiting. In any case, the equilibrium characteristics of the support and solute have a direct bearing on column performance.

Table II. Selected Equilibrium Expressions

Relationship	Application	Ref.
1. Linear $n = kc$	When concentrations are very low, or when linear adsorption occurs.	25
2. Langmuir $n = \frac{k_a c}{1 + k_b c}$	Popular isotherm in chromatographic and adsorption systems. This isotherm is asymptotically linear when c is very small or very large.	27
3. Freundlich $n = k_a c^{1/k_b}$	Empirical isotherm which can be made to fit most equilibrium data, but has no asymptotic limits.	27

Table III. Selected Mass Transfer Expressions

Relationship	Application	Ref.
1. $\frac{\partial n}{\partial t} = k_{ma} c - k_{mb} n$	Mass transfer is finite.	25
2. $\frac{\partial n}{\partial t} = k_{ma} (c - c_{eq})$	Mass transfer is finite and approaches equilibrium: Special case of previous relationship.	25
3. $\frac{\partial n}{\partial t} = D_m [\frac{\partial^2 n}{\partial r^2} + \frac{2}{r} \frac{\partial n}{\partial r}]$	Mass transfer is finite, and solute diffuses in and out of a sphere. Additional boundary conditions reflecting particle environment are required here.	21

Prediction of Elution Profiles (Linear Equilibrium). For the case of local linear equilibrium (infinite rate of mass transfer), Lapidus and Amundson (25) derived equations for computing concentration distributions in a packed column. With concentrations at the inlet of the column, and initial conditions throughout the column known, concentration profiles at a specific distance from the column inlet can be computed. The derivation was based on a semi-infinite column, which differs mathematically from a finite column, in that effects of the mobile phase leaving the stationary phase are not modeled. Nonetheless, the solution obtained is useful for giving a qualitative picture of important parameters in column performance. The equation is:

$$c(x,t) = [f_1(x,t) + f_2(x,t)] \exp(\frac{vx}{2D} - \frac{v^2 t}{4D\gamma}) \quad [12]$$

where

$$f_1(x,t) = \frac{1}{2}\sqrt{\frac{\gamma}{tD\pi}} \int_0^\infty c_{init.}(s) [\exp(\frac{-vs}{2D})][\exp(\frac{-\gamma(s-x)^2}{4Dt}) - \exp(\frac{-\gamma(s+x)^2}{4Dt})] ds \quad [13]$$

and

$$f_2(x,t) = \frac{x}{2}\sqrt{\frac{\gamma}{\pi D}} \int_0^t c_o(s) \exp[\frac{v^2 s}{4D\gamma} - \frac{\gamma x^2}{4D(t-s)}] \frac{ds}{(t-s)^{3/2}} \quad [14]$$

where

$$\gamma = 1 + \frac{K_i}{\alpha} = 1 + k'_i$$

K_i = distribution coefficient

$$\alpha = \frac{V_M}{\epsilon V_s}$$

$c_{init.}(s)$ = initial concentration profile in the column

$c_o(s)$ = inlet concentration profile for the column

Note that D, the dispersion coefficient, is not the molecular diffusivity, but a measure of combined dispersive effects inherent in packed bed operations, of which molecular diffusivity is a minor component. As pointed out by Kramers and Alberda (24), eddy diffusivity involves fluctuations of a statistical nature, and should not be applied to macroscopic effects, such as by-passing and mixing. This equation is important because it allows the modeling of chromatographic results using the dispersion coefficient as a free parameter.

Chromatographic zone broadening, however, is also influenced by finite mass transfer rates, or porous diffusion. Selection of the proper relationship between n and c is therefore needed.

The functions $f_1(x,t)$ and $f_2(x,t)$ allow solutions to be derived for special cases of inlet and initial conditions. Note that s is a dummy variable of integration. In the case where the column has been washed with eluent and a sample has not been introduced, initial condition 3 from Table I applies, and $f_1(x,t) = 0$. After a sample has been introduced and washed into the column by eluent, boundary condition 1 from Table I with $c_a = 0$ applies and $f_2(x,t) = 0$. The eluent volume is $Q = Avt\alpha$. The two limiting cases mentioned above may be superimposed, offset from one another by $A v \alpha t' = Q' = V_F$ where Q' is the feed volume. This superimposition is applicable only in the case of linear equilibrium, which yields symmetric solutions. Figure 4 shows the results of these equations graphically for selected values. These values may be particularly applicable to proteins in size exclusion supports.

In size exclusion chromatography, components are excluded from the resin on the basis of size. By definition, solutes are not absorbed. These supports have a linear equilibrium, since a constant volume within the resin is available for a given size solute. When the support comes to equilibrium with the mobile phase, the volume in the pores available to the solute will have the same molar concentrations as the surrounding mobile phase.

$$\frac{c_i'}{V_m} = \frac{n_i'}{K_D \epsilon V_s} \qquad [15]$$

where K_D is the fraction of the pore volume available to the solute, and is independent of molar concentrations. For this special case, the familiar expression

$$k_i' = \frac{n_i'}{c_i'} = K_D \frac{\epsilon V_s}{V_m} \qquad [16]$$

can be derived. Hence K_D is analogous to K_i', the distribution coefficient as defined previously where K_D addresses the special case for size exclusion being the only partitioning factor. For a solute which is completely excluded, $K_D = 0$, $k_i' = 0$ and $\gamma = 1$. For a solute which is completely included, $K_D = 1$, $k_i' = \frac{\epsilon V_s}{V_m} = \frac{1}{\alpha}$, and $\gamma = 1 + \frac{1}{\alpha}$.

Estimating Maximum Sample Volume. The maximum amount of sample volume which may be introduced into the column at any one time is readily defined for size exclusion supports. Since these supports exhibit no adsorption, their capacity for a solute may range from 0 (totally excluded solute) to $\frac{\epsilon V_s}{V_m}$ ($\frac{1}{\alpha}$, the ratio of support pore volume to mobile phase volume). A solute moves through the column in a time equal to the total volume available to that solute within the column, divided by the volumetric flow rate of eluent. Since all solutes in the column experience the same volumetric flow rate, it is the difference in column volumes available to the different solutes which defines the separation. For a binary system, the two elution volumes are:

$$V_{E_1} = V_m + K_1' \epsilon V_s \qquad [17]$$

(for component 1 which eluted first)

and

$$V_{E_2} = V_m + K_2' \epsilon V_s$$

(for component 2 which elutes second).

Since the difference in elution volumes reflects the extent of separation, the smaller elution volume, plus the volume of the feed, can be no larger than the elution volume of the second peak, V_{E_2},

$$V_{E_1} + V_F \leq V_{E_2} \qquad [18]$$

Otherwise the two peaks will partially overlap each other. Hence, it can be said that $V_F \leq V_{E_2} - V_{E_1} = \Delta V_E$. The difference in elution volumes is further given by:

$$V_{Fmax} = \Delta V_E = K_2' \epsilon V_s - K_1' \epsilon V_s$$

$$= (K_2' - K_1') \epsilon V_s \qquad [19]$$

and since, for size exclusion,

$$0 \leq K' \leq 1$$

the largest difference between elution volumes is ΔV_{Emax}:

$$\Delta V_{Emax} = \epsilon V_s$$

Similarly, this can be related to capacity factors for size exclusion chromatography:

$$(K_2' - K_1') \epsilon V_s = (k_2' - k_1') V_m$$

where:

$$0 \leq k' \leq \frac{\epsilon V_s}{V_m}$$

which gives

$$[(K_2' - K_1') \epsilon V_s]_{max} = \epsilon V_s \qquad [20]$$

since $K_2' = 1$, $K_1' = 0$ for $k' = \epsilon V_s / V_m$ and $k' = 0$, respectively. This result would imply that increasing support pore volume may allow greater feed volumes. This is true for columns of identical length. Increasing length alone does not, in general, increase throughput, since concomitant increases in elution time, and thus, processing time will also occur for constant eluent velocity, v. Column throughput may be increased by increasing column radius within the constraints mentioned previously (see section on Column Area).

Electrophoresis

Background. There have been many applications of an electric potential to separation processes. Electrophoresis is a separations process of great resolving power, capable of 10^6 theoretical plates per meter, according to Jorgenson & Lukacs (28). Applications of electric potential to different flow systems have accelerated in recent years. It is important, in the development and analysis of these systems, to understand the nature of an electrolyte system, and the events which occur when such a system is placed under an electric field.

Factors Affecting Electrophoretic Mobilities. The rate of migration of a solute through a conducting liquid in an electric field is based primarily upon three physical phenomena. These phenomena are: the presence of a charge on the particle, hydrodynamic drag around the solute, and relaxation forces in the ionic atmosphere of the solute. These forces are discussed by Overbeek and Bijsterbosch (29) in an excellent review. There are also three macroscopic affects which may occur in electric environments which affect the net migration of a solute. These are: electro-osmosis, (30), natural convection due to Joule heating, and forced convection, in the case of an elution process.

All charged particles or solutes in free solution attract ions of the opposite sign. These attracted ions cluster around the outer surface of the solute, creating an ionic environment which is equal and opposite to the particle's charge. These charged layers are known as the electric double layer, and have been studied in the context of electrophoresis by many investigators (31, 32, 33). Under the influence of an electric potential, charged species are attracted in one direction at magnitudes relative to their zeta potentials. The zeta potential is the electric potential of the solute at the surface of shear, the boundary between the moving particle's associated counterion layer and the bulk. The zeta potential, in general, is of the same sign but of a smaller magnitude than the solute charge found by titration (29). The dimension of the double layer is important in the theoretical analysis of electro-induced separations, because it is a measure of how closely the counterions surround the solute. Several developments of the impact of the double layer dimension on electrophoretic mobility have been published for a variety of geometries, and have been reviewed elsewhere (29).

Hydrodynamic drag causes a resistance to the flow of a particle at its outer surface. In general, particles and large biological molecules obey some form of Stokes law for drag in creep flow, although modifications for geometry are sometimes required. The critical dimension in all cases is the distance from the particle center into the double layer at which flow and hence, drag, are actually occurring. This is referred to as the "slipping plane" or "surface of shear" (29). Although the exact location of the "slipping plane" is not explicitly known, it is in the region of the "Inner Helmholz Plane" or "Stern Layer". The Helmholtz plane distinguishes between counterions that are physically absorbed to the particle, compared to those surrounding it in a boundary layer extending into the bulk fluid.

All charged particles migrate under the influence of electric potentials. This applies to the counterions directly surrounding the solute. Since these ions are opposite in charge to the solute, they migrate in the opposite direction, exerting an extra drag on the particle which is not predicted merely by hydrodynamic drag. This effect is greatest when the counterion layer thickness is large compared to the characteristic dimension of the solute.

When a particle migrates in a direction opposite to its counterion layer (i.e. when a potential is applied), a distortion of the ionic boundary layer of the particle takes place. The particle will no longer be in the center of its double layer, but rather, towards the edge to which it is migrating. This polarization results in the build-up of opposite charge in the opposite direction of the applied potentials. This segregation of charge relative to the solute results in a small attraction in the direction opposite to the applied potential, thereby reducing mobility (Figure 5). This is known as a relaxation effect and may reduce mobilities by 10-50% (29). This phenomenon is greatest when the dimension of the counterlayer is on the order of the dimension of the solute.

Several microscopic effects impact the electrophoretic mobility, μ, of a solute. The electrophoretic mobility is defined as the velocity with which a particle moves in a field divided by the strength of the field, and has units of length squared per time per voltage. It can be seen, by the complexity of these factors, that the electrophoretic mobility is dependent on solute size and charge, as well as medium viscosity, ionic strength and temperature.

Electroosmosis. Electroosmosis occurs in systems with applied potentials and results from preferential adsorption of charges at a fixed surface, such as a column wall. This ionic adsorption results in the build-up of a charged "counter" layer at the surface, which migrates electrophoretically. This flux along the wall induces a convective flux in the bulk due to viscous shear. If

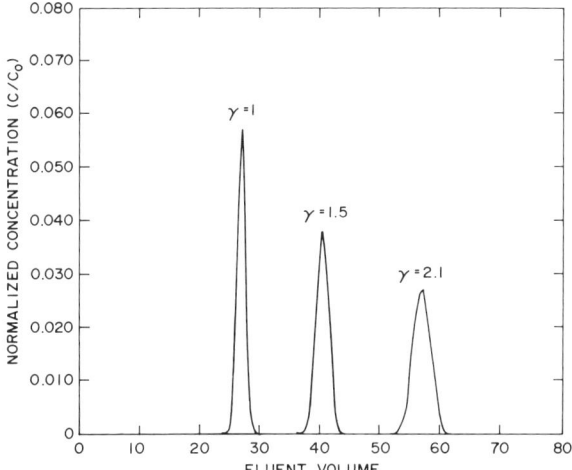

Figure 4. Calculated elution profile to simulate a protein separation on Sephadex G-75 using equation (12). This simulation is based on parameter values for a 30 x 1.5 cm column with a flowrate of 1.77 ml/min and a longitudinal diffusivity of 0.01 cm^2/min. The ratio of mobile phase volume to pore volume was 0.9, and the sample volume was 0.17 ml. Capacity factors for each of the solutes are 0, 0.5, and 1.1, respectively.

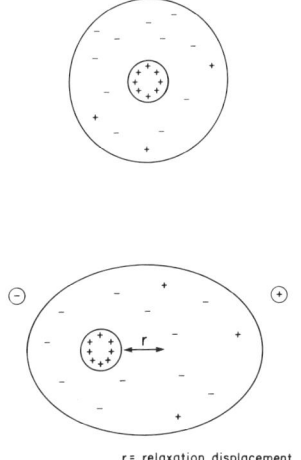

Figure 5. Schematic representation of the relaxation effect for a charged solute experiencing an electric potential.

the system is closed, and the fluid is incompressible, conservation of mass requires that the net flow across any plane be zero (30). Therefore, a circulating flow occurs, with flow in the center of the cell occurring in a direction opposite to the flow at the walls. Obviously, if a sample is introduced uniformly across a cell, the shape of the sample band will be distorted by the electroosmotic flux. Therefore, either resolution is decreased, or the amount of the channel which may be used is limited to portions in which the fluid velocity is nearly equal. Electroosmosis has been extensively reviewed (30).

There have been three methods used to minimize electro-osmosis, or its effect on separation. The first is to use an inert coating at the cell walls which will not adsorb ions. One such coating is methylcellulose (30). The problem with coatings of this nature, however, is that they are difficult to attach to glass. Several coatings have been successfully employed in solving this problem including one used in tests during an Apollo-Soyuz flight (34), and more recently, on the Space Shuttle (35).

Lowering the surface to volume ratio of a cell is another method of reducing the impact of electroosmosis on separations. Widening the apparatus will accomplish this. It has been shown (30,34) that rectangular channels are much more suitable for this kind of manipulation than circular channels. The disadvantage of widening a channel lies in reduction of the surface area/volume ratio, and hence, reduced heat dissipation. At the same time, power requirements are increased, due to the cell's wider cross section.

The last method of reducing the impact of electro-osmosis is with porous media. Resins and gels dampen out non-ideal flow effects and can minimize a nonuniform flow profile. The smaller the resin, or tighter the cross-linking, the more effective the dampening. These types of supports may have other properties, such as size exclusion properties, or the ability to adsorb charges, which are not always desirable.

Joule Heating, Viscosity and Buoyancy. When an electrical current is passed through a medium, energy is converted due to the resistivity of that medium, and is given off as Joule heat. When the current is large, the heating is great. Since most cells exchange heat at one or more boundaries, a temperature gradient arises in the cell. A thinner cell minimizes this effect. The effects of gravity on bodies of different density produce a naturally convective (and macroscopic) flow. Hot, lighter bodies (usually in the center of the cell) rise, while other bodies sink. It is the elimination of this effect which makes electrophoresis in microgravity so appealing. Different temperatures also create different liquid viscosities. Since the viscosity of the liquid impacts hydrodynamic drag, temperature gradients can alter electrophoretic mobilities, independent of gravitational forces. This viscosity effect is estimated to change mobilities by about 2% per degree centigrade (28).

Electrophoresis with Controlled Convection. The study of how electrophoresis within controlled convection can be applied efficiently to flow systems in preparative and industrial scale separations is an area for further development. The potential for such separations is great, based on the number and diversity of systems which have been proposed over the last fifteen years (34-50).

An electrochromatography system which is receiving attention is that of O'Farrell (40). This system (Figure 6) consists of a layered bed of size exclusion gels, in which a less porous support rests upon a more porous gel. Since the more porous gel has more pore volume available to solute, a solute's velocity in this layer is slower than in the less porous layer. O'Farrell found that an electric field could be selected, which would focus one solute at the interface of the layered gels. The solute must concentrate in one of the two layers near the interface. The solute has been shown to concentrate in the lower gel, because the local increase in electrolyte concentration decreases the local resistance, and therefore decreases the local potential. In regions other than the concentrated solute region, the net flux of solute is always toward the concentrated solute region. This system operates in batch mode for the processing of one solute, with the remaining components being eluted in the bulk.

Very recently, Gassmann, Kuo, and Zare (44) demonstrated the "electrokinetic" separation of racemic mixtures of amino acids. Their system used a potential applied along the axis of a capillary column, with open electrode reservoirs. Their results show a chromatographic type separation of dansyl derivalized L and D amino acids. The flow in this system is due to electroosmosis, and is large enough in magnitude to elute all species (positive, negative, and neutral) from the cathodic end of the column. Since the electrode reservoirs are open, buffer flows freely through the capillary, and none of the distortion associated with electroosmosis on closed systems is observed. The investigators employed a specially designed laser fluorescence detector to quantify extremely small quantities (i.e. 10^{-15} M) of solute. Although the column was 75 cm in length, analysis time was reportedly 6 to 10 minutes.

As a result of experiments originally conducted on Apollo 16 (34), commercial electrophoretic efforts have been undertaken on space shuttle flights (45,46). The apparatus involved employs free flow electrophoresis of a protein mixture for the purification and recovery of a very high valued product. The process itself is proprietary. However, data and models are available from earlier flights and other experiments (34,47,48).

Isoelectric Focusing. Isoelectric focusing has also been scaled up to take advantage of its equilibrium resolution. By equilibrium resolution, it is meant that the system depends upon solutes arriving at an equilibrium position, and that neither duration of processing, nor length of equipment, will disrupt this position. Isoelectric focusing takes advantage of the fact that the zeta potential of a protein changes according to the pH of the environment. When a protein reaches the point, in a pH gradient, where its zeta potential reaches zero (its "isoelectric point"), it ends its migration and "focuses". The focusing is due to the fact that, if the protein randomly migrates in either direction from its isoelectric point, it acquires a charge and migrates back. Bier et. al (49,50) have exploited this phenomena in a preparative recycle instrument,

140   SEPARATION, RECOVERY, AND PURIFICATION IN BIOTECHNOLOGY

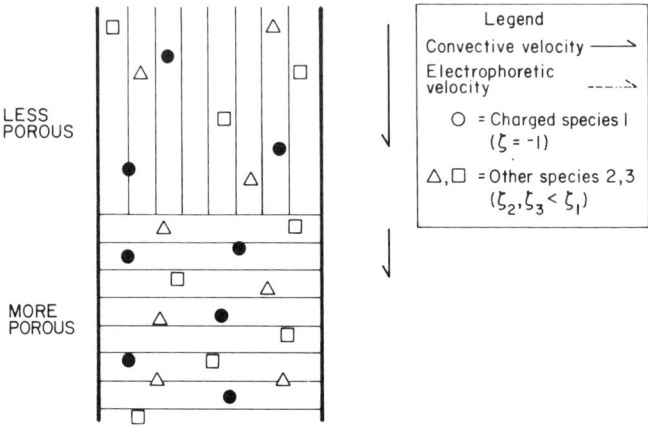

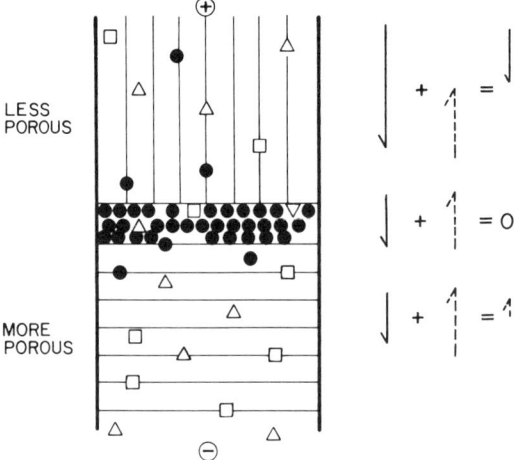

Figure 6. Schematic diagram of electrochromatographic apparatus described by O'Farrell (40).

which we estimate to be capable of processing samples at the gram level or higher. This instrument appears adaptable for a variety of capacities.

Plate Height and Resolution Concepts in Electrophoresis

Theoretical plates can be used to evaluate electrophoretic and electrochromatographic performance, in a manner analogous to chromatographic performance. These type of numerical evaluations are useful to compare results, both between different applications of the same apparatus, and between similar applications of different apparatus.

Following the definitions and derivation of rate of generation of variance with apparatus length ($d\sigma_L^2/dX$) set forth by Giddings (51), Jorgenson and Lukacs (28) defined σ_L^2 as

$$\sigma_L^2 = 2Dt$$

It should be noted here that D is most properly referred to as longitudinal dispersion, which may be attributed to a number of effects, the least of which is usually molecular diffusion. Other dispersive effects are usually lumped into "eddy diffusivity". The development of Jorgenson and Lukacs was applied to a system in which molecular diffusivity was the major cause of dispersion, but their approach appears to be applicable to other systems as well. Substituting in analysis time, $t = \dfrac{L^2}{\mu_i V}$ we obtain:

$$\sigma_L^2 = 2DL^2/\mu_i V \qquad [21]$$

where

$$\mu = \text{electrophoretic mobility} \quad \left(\dfrac{cm^2}{Volt\ min}\right)$$

so

$$N_i = \dfrac{L^2}{\sigma_L^2} = \dfrac{\mu_i V}{2D} \qquad [22]$$

In this case, the number of theoretical plates, N, is independent of length of the apparatus. Since μ is a molecular parameter, and D is related to the apparatus and the solute, the easiest variable to manipulate is the applied voltage, V. N can therefore be increased by increasing the applied voltage. Since analysis time

$$t = \frac{L^2}{\mu V} \qquad [23]$$

is inversely proportional to voltage, separating efficiency (N/t) will be maximized at high voltage and short apparatus lengths. This has been shown experimentally (28).

The plate height (HETP) is then shown to be

$$H = \frac{2DL}{\mu V} \qquad [24]$$

which is a direct function of length. This is in direct contrast to chromatography, where N is a function of length, and H is independent of length. An interesting variation may be seen by applying Ohm's law ($V = IR$) to the equations for N and H.

$$N = \frac{\mu IR}{2D} \qquad [25]$$

$$H = \frac{2DL}{\mu IR} \qquad [26]$$

Joule heating is proportional to $W = I^2 R$. Hence, a continuous phase having a high R would serve to increase both applied voltage and heat generation at constant current. Assuming good control of heat removal a low ionic strength buffer would thus be most appropriate for high separation efficiency. Jorgenson and Lukacs also suggest that the sample be of an ionic strength less than that of the buffer, in order to negate adsorption and interaction affects within the sample that may lead to zone broadening.

Jorgenson and Lukacs also described resolution in electrophoresis systems as

$$R_s = \frac{1}{4} [\frac{\bar{\mu} V}{2D}]^{\frac{1}{2}} [\frac{\mu_1 - \mu_2}{\bar{\mu}}]$$

or

$$R_s = 0.177 (\mu_1 - \mu_2) (\frac{V}{\bar{\mu} D})^{\frac{1}{2}}$$

We see here that high voltage also enhances resolution. If another flux, for example, a convective flux, is imposed upon the system, we speculate that this equation could be empirically rewritten as:

$$R_s = 0.177\,(\mu_1 - \mu_2)V\,\left(\frac{1}{(\bar{\mu}V + vL)D}\right)^{\frac{1}{2}} \quad [27]$$

where

$$v = \text{convective flux}$$

and

$$\bar{\mu} = \frac{\mu_1 + \mu_2}{2}$$

As pointed out by Jorgenson and Lukacs, this allows another variable for optimization of resolution. If $\bar{\mu}V + vL$ is made very small, i.e. $\bar{\mu}V \approx -vL$, resolution may be greatly increased. In order for both solutes to elute from the same end of the apparatus, however, $-vL < \mu_2 V$. The convective flux must be large enough to push both solutes through the apparatus. Thus, for example, if a solute i were to migrate in a direction opposite to flow due to its inherent electrophoretic mobility, the flux given by vL would have to be greater in magnitude than the electrophoretic flux, $\mu_2 V$. Hence, the condition of $-vL < \mu_2 V$ is proposed if all solutes are to elute in the direction of the flow. Analysis time is given by:

$$t = \frac{L^2}{\mu V + vL} \quad [28]$$

Since the variables v and V may be varied independently, it is seen that length of apparatus is not necessarily a constraining factor, as it is in liquid chromatography. Hence separation optima may be discerned which are independent of length.

Theoretical Considerations in Electrokinetic Separations. Models have been proposed for electrophoretic (52) and electrochromatographic systems. Similar to developments in chromatography, we may write a model equation for these systems:

$$D\frac{\partial^2 c}{\partial x^2} = v\frac{\partial c}{\partial x} + \mu\frac{\partial(Ec)}{\partial x} + \frac{\partial c}{\partial t} + \frac{1}{\alpha}\frac{\partial n}{\partial t} \quad [29]$$

where the new term $\mu\frac{\partial(Ec)}{\partial x}$ = flux due to electric potential. This equation may be solved with the same boundary conditions which are applied to chromatographic processes. A similar equilibrium or mass transfer equation may be used as well. We find, however, that another relation becomes necessary for μE in terms of c. If we assume that μE is independent of concentration (53) or that concentration in the column is nearly constant, we speculate that the solution of Lapidus and Amundson (25) may be adapted, where v is replaced everywhere by $v + \mu E$. Results from the substitution are shown in Figure 7, which shows a numerical solution for equations (12) to (15) modified by replacing v with $v + \mu E$.

This calculation shows how resolution might be obtained for a three component mixture, where two components have the same liquid chromatographic capacity factors ($k_i = 0.5$, $i = 2, 3$) with slightly different electrophoretic fluxes ($\mu_3 E = -2.7$; $\mu_2 E = -2.2$).

The assumption that the electric field is homogeneous throughout the system is only a first approximation. Many electrophoretic systems, such as isotachophoresis, rely upon potentials varying with the local ionic strength to affect a separation. In modeling these systems, one must find a function which describes the local potential in terms of total concentration. The most commonly used form is Poisson's Law:

$$\frac{\partial E}{\partial x} = \frac{F}{\varepsilon \varepsilon_o} (\sum_i z_i^+ c_i^+ + \sum_j z_j^- c_j^-) \qquad [30]$$

where

ε = dialectric constant of electrolyte solution
ε_o = permittivity of free space
F = Faraday's constant
z = valence

This relationship was used effectively by Coxon and Binder (54), and Lim and Franses (55) to solve electrophoretic systems in which flow was not applied. Lim and Franses used ionic reactions at the electrodes as boundary conditions to model apparently decreasing electrophoretic mobilities in an electrophoretic mass transport analysis. They showed that increasing ionic strength at one electrode decreased the local potential, thus decreasing mobility. Coxon and Binder specified concentrations at electrodes to arrive at a model for ionic interfaces in isotachophoretic processes. Each team investigated different systems, thus resulting in different boundary conditions.

These equations show that the final elution profiles will be affected by several parameters, such as dispersivity, field strength, eluent velocity, as well as time, distance, and ionic strength distribution within the column.

There have been numerous advances in both the understanding and applications of electric potentials to separations in recent years. Electrophoresis has been met with skepticism as an industrial unit operation (56,57), but recent applications would suggest a "rebirth" of interest in the potential of these systems, especially in biological applications.

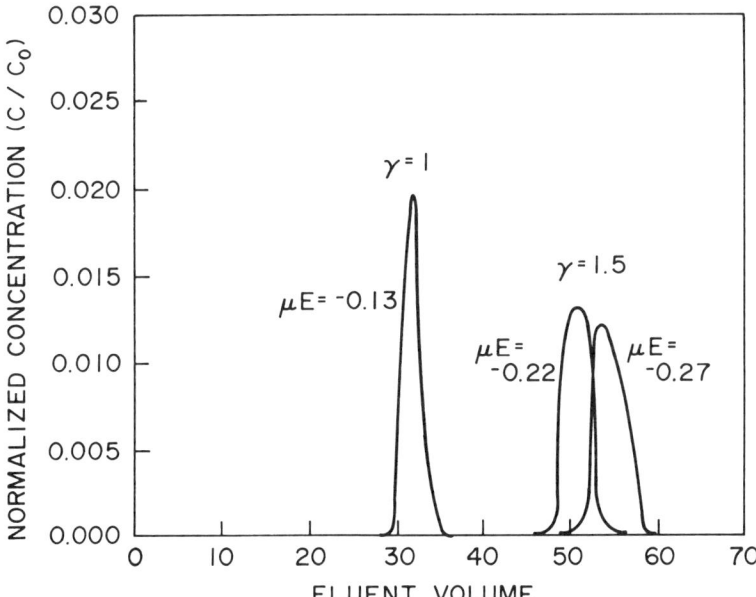

Figure 7. Calculated elution profile to simulate a protein separation on Sephadex G-75 using a modified version of equation (12). This simulation is based on parameter values for a 30 x 1.5 cm. column with a flow rate of 1.77 ml/min, a longitudinal diffusivity of 0.01 cm^2/min, and an electric field strength of -25 volts/cm. The ratio of mobile phase volume to pore volume is 0.9, and the sample volume is 0.17 ml. Capacity factors for each of the solutes are 0, 0.5 and 0.5. The electric potential results in an electric flux term of -0.13, -0.22, and -0.27 cm/min for each solute respectively, in order of elution. These values correspond to electrophoretic mobilities of 8.6 x 10^{-5}, 14.7 x 10^{-5} and 18.0 x 10^{-5} $cm^2/V\cdot s$, respectively.

Summary

The practical use of chromatographic and electrophoretic separations in the biotechnology industry will be aided by correlations capable of relating bench scale results to process scale conditions. Published physical and chemical property data on chromatographic systems for engineering calculation purposes currently appears to be limited. Hence, an effort was made in this chapter to present equations which can be used with physical parameters for which values have been reported or which are readily determined on a bench scale. Since size exclusion chromatography is important in fractionation of proteins, the rationale and equations were presented for estimating:

1. maximum sample volume (equations (19) and (20));
2. elution profiles for non-interfering components (equation (12) to (14)) with results for an example given in Figure 4; and
3. changes in column length if particle size of a chromatographic support is changed upon scale-up (equation (6)).

These calculations require knowledge of column void fraction ($\bar{\alpha}$); fractional pore volume of the stationary phase (ϵ); capacity factors (k_i) or distribution coefficients (K_i); dispersitivity or diffusivity of the solute (D); and initial and inlet column conditions. The determination of k_i' is described in reference 5. Knowledge of limits for sample volume, elution volume, and column length for a given target of product purity and recovery allows a "best-case" estimate of throughput, eluent volumes, and support costs using correlations described previously (5).

Electrophoresis gives fantastic resolution of small protein and peptide samples on a bench scale, and has prompted numerous experimental approaches for enhancing this technique. Recent developments include the systems of O'Farrell (40), Gassmann et.al. (44), and the Space Shuttle (45,46). Scale-up of a preparative recycle instrument was recently reported by Bier (50).

Key parameters for describing electrophoresis are plate count (equation (22)) and resolution (equation (27)). Unlike chromatography, the number of plates is a function of applied voltage, V (equation (22)) rather than length. The combination of size exclusion chromatography with an electrophoretic driving force is proposed for continuous elution and separation of a multicomponent mixture. An estimate of profiles based on an unsteady material balance (equation (29)) and equations ((12) to (14)) suggests such a separation is theoretically possible (Figure 7). If so, a compact and scalable unit operation combining convective and electrical effects to give rapid and extraordinary resolution of proteins in existing size exclusion chromatographic systems should be possible.

Acknowledgments

The material in this work was supported by NSF Grant CPE 8351916 and Kraft, Inc. Support for activities in process chromatography and scale-up from Artisan Industries is also acknowledged. We also thank Mr. K. Ruettimann for helpful comments on electrophoretic effects.

Nomenclature

a = empirical constant

A = column cross sectional area (cm^2)

c = concentration in mobile phase (*moles / l*)

c_a = concentration constant (*moles / l*)

c_{eq} = concentration in mobile phase if mobile phase is in equilibrium with stationary phase (*moles / l*)

c_i = solute concentration in mobile phase (*moles / l*)

c_i' = moles of solute in mobile phase (*moles*)

c_{init} = initial conditions for column as a function of distance, or a constant (*moles / l*)

c_o = concentration at column inlet as a function of time, or a constant (*moles / l*)

D_i = dispersivity associated with solute i ($\frac{cm^2}{min}$)

D_m = diffusivity inside particle ($\frac{cm^2}{min}$)

d_p = support particle diameter (μm)

$d_{p,A}$ = original scale particle diameter (μm)

$d_{p,x}$ = scaled up support particle diameter (μm)

E = electric field strength (*Volts/cm*)

F = Faraday's constant

H = height equivalent to a theoretical plate (cm)

I = applied current ($amps$)

k, k_a, k_b = equilibrium constants

K_D = fraction of pore volume available to a particular solute

k_i' = capacity factor ($\frac{moles}{moles}$)

k' = average of two capacity factors $\frac{k_1' + k_2'}{2}$

K_i = pore volume distribution coefficient ($\frac{moles}{l} / \frac{moles}{l}$)

K_i' = support volume distribution coefficient ($\frac{moles}{l} / \frac{moles}{l}$)

k_{ma}, k_{mb} = mass transfer constants ($\frac{1}{time}$)

L = length of apparatus (cm)

L_A = original scale apparatus length (cm)

L_x = scale up apparatus length (cm)

m = empirical constant

n = concentration in stationary phase ($moles/l$)

n_i = solute concentration in the stationary phase ($\frac{moles}{l}$)

n_i' = moles of solute in stationary phase ($moles$)

N = average of two plate counts, $\frac{N_1 + N_2}{2}$

N_i = number of theoretical plates based on solute i

Q = volume of flow (ml)

Q' = feed volume (ml)

r = distance along the support particle radius (μm)

R	=	continuous phase resistance (*ohms*)
R_s	=	resolution
s	=	dummy variable of integration
t	=	time
t'	=	time of feed duration
v	=	mobile phase linear velocity (*cm / min*)
V	=	applied voltage (*volts*)
V_{Ei}	=	elution volume of solute i (*ml*)
ΔV_E	=	difference in elution volumes of two solutes (*ml*)
ΔV_{Emax}	=	maximum difference in elution volumes (*ml*)
V_F	=	feed volume (*ml*)
V_{Fmax}	=	maximum sample volume (*ml*)
V_m	=	mobile phase volume (*ml*)
V_{pores}	=	volume of support pores (*ml*)
V_s	=	stationary phase volume (*ml*)
V_T	=	total column volume (ml)
W	=	power (*watt, volt-amps*)
x	=	longitudinal distance from column inlet (*cm*)
z_i	=	valence of solute i

Greek

α = ratio of mobile phase volume to stationary phase pore volume ($\frac{V_m}{\epsilon V_s}$)

$\bar{\alpha}$ = column void fraction ($\frac{V_m}{V_T}$)

ε = pore volume fraction ($\frac{V_{pore}}{V_S}$)

ε = dielectric constant of media

ϵ_o = relative permittivity

$\gamma = 1 + k_i'$

μ_i = electrophoretic mobility of solute i ($\frac{cm^2}{volt \cdot min}$)

$\bar{\mu}$ = average of two electrophoretic mobilities, $\frac{\mu_1 + \mu_2}{2}$

ϕ = ratio of capacity factors for two solutes, $\frac{k_2'}{k_1'}$

σ = standard deviation

σ_L^2 = variance

ζ_i = zeta potential of solute i

Literature Cited

1. Scott, C.D.; Spence, R.D.; Sisson, W.G. *J. Chromatogr.* 1976, 126, 381.
2. Canon, R.M.; Beagovich, J.M.; Sisson, W.G. *J. Chromatogr.* 1980, 15(3), 655.
3. Beagovich, J.M.; Sisson, W.G. *Resources and Conservation.* 1982, 9, 219.
4. Beagovich, J.M.; Byers, C.H.; Sisson, W.G. *Separation Science and Tech.* 1983, 18(12 & 13), 1167.
5. Ladisch, M.R.; Voloch, M.; Jacobson, B. *Biotechnol. Bioeng. Symp. No. 14.* 1984, 525.
6. Belter, P.A. "Recovery Processes-Past, Present and Future," 1984th National Meeting of the American Chemical Society, 1982.
7. Sherwood, T.K.; Pigford, R.L.; Wilke, C.R. *Mass Transfer.* 1975, 548-592.
8. Treybal, R.E. *Mass Transfer Operations.* McGraw Hill, N.Y. 1968, 491-568.
9. Tudge, A.P. *Can. J. Phys.* 1961, 39, 1600.
10. Goldstein, S. *I. Proc. Roy. Soc.* 1953, A, 219, 151.
11. Absolom, D.R., *Sep. Purif. Meth.* 1981, 10(2), 239.
12. Ladisch, M.R.; Huebner, A.L.; Tsao, G.T. *J. Chromatogr.* 1978, 147, 185.

13. Ladisch, M.R.; Tsao, G.T. *J. Chromatogr.* 1978, 166, 85.
14. Newman, R.N.; Rudge, J.R. and Ladisch, M.R. *Reactive Polymer J.,* submitted 1986.
15. Pieri, G.; Piccerdi, P.; Muratori, G.; Cavalini, L. *La Chem e L'Ind.* 1983 65(5) 331.
16. Barford, R.A.; McGraw, R.; Rothbart, H.L. *J. Chromatogr.* 1978, 166, 365.
17. Kucera, E. *J. Chromatogr.* 1965, 19, 237
18. Martin, R.L.M.; Synge, A.J.P. *Biochem. J.* 1941, 35, 1358.
19. Snyder, L.R.; Kirkland, J.J. *Introduction to Modern Chromatography.* 2nd ed. 1979.
20. Aris, R.; Amundson, N.R. *AIChE Jour.* 1957, 3(2), 280.
21. Babcock, R.E.; Green, D.W.; Perry, R.H. *AIChE Jour.* 1966, 12(5), 922.
22. Brenner, H. *Chem. Eng. Sci.* 1962, 17, 229.
23. Danckwert, P.V. *Chem. Eng. Sci.* 1953, 2(1), 1.
24. Kramers, H.; Alberda, G. *Chem. Eng. Sci.* 195U3, 2, 173.
25. Lapidus, L.; Amundson, N.R. *J. Phys. Chem.* 1952, 56, 984.
26. Pearson, J.R.A. *Chem. Eng. Sci.* 1959, 10, 281.
27. Wankat, P.C.; *Large Scale Adsorption and Chromatography,* CRC Press, Boca Raton, FL, in press. 1986
28. Jorgenson, J.W.; Lukacs, K.D. *Anal. Chem.* 1981 (53) 1298.
29. Overbeek, J.Th.G.; Bijsterbosch, B.H. *Electrokinetic Separation Methods.* 1979, 1-32.
30. Vanderhoff, J.W.; Micole, F.J. *Electrokinetic Separation Methods.* Elsevier, Holland. 1979, 81-93.
31. Verwey, E.J.W.; Overbeek, J.Th.G. *Theory of the Stability of Lyophobic Colloids.* 1948, 22-46.
32. Stern, O. *Z. Electroc.* 1924, 30, 508.
33. Sparnaay, M.J. *The Electrical Double Layers* Pergamon Press, Great Britain, 1972, 46-61.
34. Vanderhoff, J.W.; Micole, F.J.; Krumrine, P.H. *Electrokinetic Separation Methods.* 1979, 121-141.
35. Chao, J.F.; Holleln, H.C.; Huang, C-R. *Ind. Eng. Chem. Fundam.* 1985 24, 489.
36. Eisinger, R.S.; Alkire, R.C. *J. Electrochem. Soc.* 1983, 130(1), 93.
37. Gobie, W.A.; Beckwith, J.B.; Ivory, C.F. *Biotechnol. Prog.* 1985, 1(1), 60.
38. Hannig, K. *Electrophoresis,* Academic Press, NY 1967, 423-472.
39. Nerenberg, S.T.; Pogojeff, G. *Am. J. of Clin. Path.* 1969, 51(6), 728.
40. O'Farrell, P.H. *Science.* 1985, 227, 1586.
41. Oren, Y.; Soffer, A. *J. Electrochem. Soc.* 1978, 125(6), 869.
42. Reis, J.F.G.; Ramkrishna, D.; Lightfoot, E.N. *AIChE Journal.* 1978, 24(4), 679.
43. Salik, J.; Roch, P. *J. Chromatogr.* 1972, 71, 459.
44. Gassmann, E.; Kuo, J.E.; Zare, R.N. *Science,* 1985 230, 813.

45. Walters, S., *Mech. Eng.,* 1982, 104, 46.
46. Burlis, N.W., "Agony and Ecstasy of EOS" 187th National Meeting of the American Chemical Society, 1985.
47. *Materials Processing in the Reduced Gravity Environment of Outer Space; Proceedings of the Annual Meeting,* 1981.
48. *Electrophoresis '82; Proceedings of the Fourth International Conference,* 1982.
49. Bier, M.; Egen, N.B.; Cellgyes, T.T.; Twitty, G.E.; Mosher, R.A. *Pepthides; Structure and Biological Functions,* Pierce Chemical Company, 1979, 79.
50. Bier, M. "Recycling Isoelectric Focusing" 187th National Meeting of the American Chemical Society. 1985.
51. Giddings, J.C. *Sep. Sci.,* 1969, 4, 181.
52. Bier, M.; Palusinski, O.A.; Mosher, R.A.; Saville, P.A. *Science.* 1983, 219, 1281.
53. Reis, J.F.G.; Lightfoot, E.N.; Lee, H-L. *AIChE Journal.* 1974, 20(2), 362.
54. Coxon, M.; Binder, M.J. *J. Chromatogr.* 1974, 95, 133.
55. Lim, K.H.; Franses, E.I. *Chem. Eng. Commun.* 1983, 22, 181.
56. Graff, G.M. *Chem. Eng.* 1983, June 13, 22.
57. Webber, D. *C & EN.* 1984, April 16, 11.

RECEIVED April 16, 1986

11

Mathematical Modeling of Bioproduct Adsorption Using Immobilized Affinity Adsorbents

Somesh C. Nigam and Henry Y. Wang

Department of Chemical Engineering, The University of Michigan, Ann Arbor, MI 48109-2136

> The use of small affinity adsorbent particles immobilized in hydrogel beads has been investigated for whole broth processing (1). The adsorbent particles can contain biospecific ligands covalently attached to a porous solid support. A mathematical model was developed to study bioproduct adsorption using immobilized affinity adsorbent beads in batch operation. The performance of immobilized and freely suspended affinity adsorbents was compared by calculating adsorption rates and selectivities for four different bead geometries. Simulation results indicate that the performance of finely ground adsorbent particles immobilized in hydrogel beads is superior compared to freely suspended adsorbents. The mathematical model was further used for simulation studies to investigate the effect of bead design parameters on product adsorption.

Affinity adsorption, due to its high degree of selectivity, offers a viable alternative to conventional crude bio-product separation schemes. However, there are several problems associated with using freely suspended affinity adsorbent particles in the whole broth. Large adsorbent particle size is required to ensure easy handling in the broth. But this leads to high internal mass transfer resistance which significantly reduces the adsorption rate. The presence of various organic macromolecules in the broth can lead to rapid fouling of the adsorbent particles. Also, the broth may contain by-products in substantial concentration which may compete with the desired product for the ligand.
The use of small affinity adsorbent particles immobilized in hydrogel beads has been proposed to circumvent some of these problems (1). The hydrogel matrix can be provided by Ca-Alginate, K-Carrageenan or any other reversible gel. Previous research in our laboratory has indicated that significantly higher adsorption rates and overall adsorption capacities can be achieved by using immobilized affinity adsorbent beads in the whole broth. These beads provide low overall internal mass transfer resistance due to the

small adsorbent particle size. A relatively large bead size (1-3 mm) ensures easy recovery from the whole broth at the end of adsorption process. Polymerization of these hydrogels can be reversed by manipulating the concentration of exogenous cations and inducing temperature shifts. Adsorbent particles with bound product can be easily recovered by dissolving away the hydrogel matrix. Large macromolecules present in the whole broth are excluded from the hydrogel because of pore size restriction. Undesired macromolecules that do penetrate foul the outer hydrogel layer first. This saves most of the ligand distributed inside the bead. Many of the available biospecific ligands used for bioseparation are more expensive compared to the product itself. Retrieving and reusing the ligands after bioseparation is crucial to the economic success of an affinity bioseparation process. Covalent attachment of the ligand to an insoluble support was used to minimize leakage. Therefore, the ligand can be re-used for subsequent bioseparations.

The purpose of this article is to formulate a model which considers simultaneous diffusion and binding reaction within the immobilized adsorbent particles. The model has been developed for batch adsorption processes. It can however be easily modified to predict product adsorption in other reactor configurations.

Theory

Affinity adsorption is a separation technique based on specific and reversible binding of two biologically active compounds. Numerous biological compounds recognize and bind to specific compounds. For example enzymes form complexes with substrates in the course of their normal catalytic mechanisms. Similarly, antibodies form very strong complexes with their respective antigens. Various proteins also interact selectively with other macromolecules.

Graves and Wu have developed a simple equilibrium model for describing affinity binding reactions (2). The binding reaction between a product and an affinity ligand covalently attached to a solid support can be represented as:

$$P + L^* \underset{K_{-1}}{\overset{K_1}{\rightleftharpoons}} P.L^* \qquad (1)$$

In the simplest case the rates of adsorption and desorption can be written as:

$$r_{ads} = K_1 [P] [L^*] \qquad (2)$$

$$r_{des} = K_{-1} [P.L^*] \qquad (3)$$

where [P] is the product concentration, [L*] is the concentration of the bound ligand and [P.L*] is the concentration of the product-ligand complex.

This yields an equilibrium constant:

$$K_{eq} = \frac{[P][L^*]}{[P.L^*]} = \frac{K_{-1}}{K_1} \qquad (4)$$

In this approach it is assumed that the product molecule binds to a single binding site on the ligand through monovalent interaction. For this mechanism, the rate of adsorption can be expressed by a relation which is first order with respect to both, product and ligand concentration (Equation 2). However, there may be circumstances where the product molecule contains more than one binding site that is recognized by the ligand. Such a multivalent interaction requires a more complex analysis (3). Most of the affinity binding reactions are characterized by very small equilibrium binding constants. We will assume the rate of adsorption (r_{ads}) to be much higher compared to the rate of desorption (r_{des}) so that the affinity binding can be considered as essentially irreversible.

Figure 1 shows a schematic diagram of an immobilized affinity adsorbent bead. Hydrogel, by virtue of its extremely high water content (>90%), offers limited diffusional resistance to the desired product. It is therefore used as an inert matrix to support relatively small adsorbent particles which otherwise cannot be readily recovered from a highly heterogenous whole broth. The reduced adsorbent particle size leads to significant decline in internal diffusional resistance which offsets any marginal increase in resistance due to the hydrogel matrix itself.

Several assumptions are made to mathematically model the immobilized adsorbent. The small adsorbent particles are assumed to be distributed uniformly inside the hydrogel bead. The external mass transfer resistance due to the boundary layer is assumed to be negligible if the bulk solution is well stirred. This assumption is supported by the experimental observations of Tanaka et al. who studied diffusion of several substrates from well stirred batch solutions into Ca-alginate gel beads (4). However, the boundary conditions can be easily modified to incorporate external diffusion effects if needed. Furthermore product diffusion in both the hydrogel and the porous adsorbent is considered to follow Fickian laws and its diffusivity in each region is assumed to be constant.

The unsteady state product and ligand material balances in the different regions can be expressed as follows.

The product mass balance in the hydrogel can be represented as:

$$\frac{D}{R^2}\frac{\partial}{\partial R}\frac{(R^2\partial C_i)}{\partial R} - \frac{(3ND_{Ai} r_o^2)}{R_o^3}\frac{\partial C_{Ai}}{\partial r}\bigg|_{r=r_o} = \frac{\varepsilon_g \partial C_i}{\partial t} \qquad (5)$$

The product mass balance in the adsorbent particles using a first order binding reaction with respect to both product and ligand is given by:

$$\frac{D_{Ai}}{r^2}\frac{\partial}{\partial r}\frac{(r^2\partial C_{Ai})}{\partial r} - K_i C_{Ai} C_1 = \frac{\varepsilon_a \partial C_{Ai}}{\partial t} \qquad (6)$$

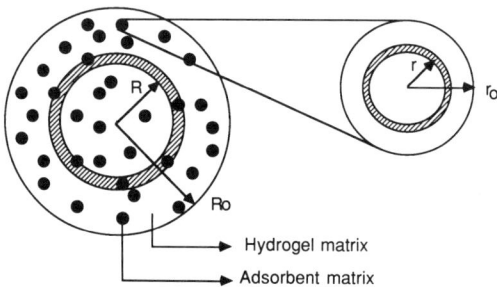

Figure 1. Schematic diagram of an immobilized affinity adsorbent bead.

Product depletion in the bulk solution of the batch adsorber is given by:

$$\frac{\partial C_{bi}}{\partial t} = \frac{-4\pi n R_o^2 D_i}{V} \left. \frac{\partial C_i}{\partial R} \right|_{R=R_o} \tag{7}$$

The ligand balance within the adsorbent particle is:

$$\frac{\partial C_1}{\partial t} = \frac{-K_i C_{Ai} C_1}{a} \tag{8}$$

In the case of negligible external diffusion resistance, the initial and boundary conditions for equations 5-8 can be written as:

Initial Conditions (at t=0):

$C_i = C_{Ai} = 0$; (zero initial product loading on the bead)

$C_1 = C_{1o}$; (uniform ligand concentration)

$C_{bi} = C_{bi}^o$; (uniform bulk concentration)

Boundary Conditions

$R = 0$: $\dfrac{\partial C_i}{\partial R} = 0$; (radial symmetry of hydrogel bead)

$R = R_0$: $C_i = C_{bi}$; (concentration continuity at hydrogel-bulk solution interface)

$r = 0$: $\dfrac{\partial C_{Ai}}{\partial r} = 0$; (radial symmetry of adsorbent particle)

$r = r_o$: $C_{Ai} = C_i$; (concentration continuity at hydrogel-adsorbent interface)

The mass balance equations given above can be represented in non-dimensional form by employing the following dimensionless variables and parameters:

$$\bar{r} = r/r_o; \quad \bar{R} = R/R_o; \quad \bar{t} = D_{ref} t/R_{ref}^2$$

$$\bar{C}_1 = C_1/C_{1o}; \quad \bar{C}_{bi} = C_{bi}/C_{bi}^o; \quad \bar{C}_i = C_i/C_{bi}^o; \quad \bar{C}_{Ai} = C_{Ai}/C_{bi}^o;$$

$$\psi_i^2 = K_i r_o^2 C_{1o}/D_{Ai}; \quad \mu_i = \varepsilon_a r_o^2 D_{ref}/R_{ref}^2 D_{Ai}; \quad \phi_i^2 = K_i R_{ref}^2 C_{bi}^o/D_{ref} a;$$

$$\gamma_i = 4\pi n R_o D_i R_{ref}^2 / V D_{ref}; \quad A_i = 3ND_{Ai} r_o / D_i R_o; \quad B_i = \varepsilon_g R_o^2 D_{ref} / R_{ref}^2 D_i$$

$$\frac{1}{\bar{R}^2} \frac{\partial}{\partial \bar{R}} \frac{(\bar{R}^2 \partial \bar{C}_i)}{\partial \bar{R}} - \frac{A_i \partial \bar{C}_{Ai}}{\partial \bar{r}} \bigg|_{\bar{r}=1} = \frac{B_i \partial \bar{C}_i}{\partial \bar{t}} \tag{9}$$

$$\frac{1}{\bar{r}^2} \frac{\partial}{\partial \bar{r}} \frac{(\bar{r}^2 \partial \bar{C}_{Ai})}{\partial \bar{r}} - \psi \, {}^2\bar{C}_{Ai}\bar{C}_1 = \frac{\mu_i \partial \bar{C}_{Ai}}{\partial \bar{t}} \tag{10}$$

$$\frac{\partial \bar{C}_{bi}}{\partial \bar{t}} = -\frac{\gamma_i \partial \bar{C}_i}{\partial \bar{R}} \bigg|_{\bar{R}=1} \tag{11}$$

$$\frac{\partial \bar{C}_1}{\partial \bar{t}} = - \sum_{i=1}^{NC} \phi_i^2 \bar{C}_{Ai} \bar{C}_1 \tag{12}$$

In the past, similar bidispersed systems have been investigated and modeled in the catalyst deactivation area (5-7). However, modeling of immobilized affinity adsorbent beads is more complex due to the non-linearity introduced by the rapid ligand binding reaction which is dependent on the product concentration.

The mathematical model described above involves non-linear, coupled, partial differential equations. The equations were solved using a Finite-Difference method. Details of this mathematical technique have been described elsewhere in the literature (8,9). Figure 2 shows a flowsheet for the numerical solution of these model equations.

Simulation Studies

Several simulation runs were carried out to gain insight into the effect of bead design parameters on the adsorption characteristics of immobilized adsorbent beads. The physical parameters (rate constant, diffusivity etc.) for the simulation studies were determined from experimental data on the adsorption of cycloheximide, a low molecular weight antibiotic, onto XAD-4 non-ionic polymeric resin (10,11) (Table I). The fit between the model and the experimentally determined adsorption curves is quite good (Figure 3).

Single component diffusion and binding. Figure 4 shows four cases which were simulated to observe the effects of immobilization in hydrogel and reduction of adsorbent particle size. Case (a) represents a freely suspended adsorbent particle of radius 1.1 mm. Case (b) represents the same size particle immobilized in a hydrogel bead of 2.8 mm. In case (c), the same adsorbent particle as in cases (a) and (b) was assumed to be crushed to 80 smaller particles which were immobilized within a hydrogel bead of radius 2.8 mm. Case (d) represents the extreme situation in which the adsorbent particle was crushed to fine powder such that the total number of particles within the immobilized bead may be regarded as infinite. This is also

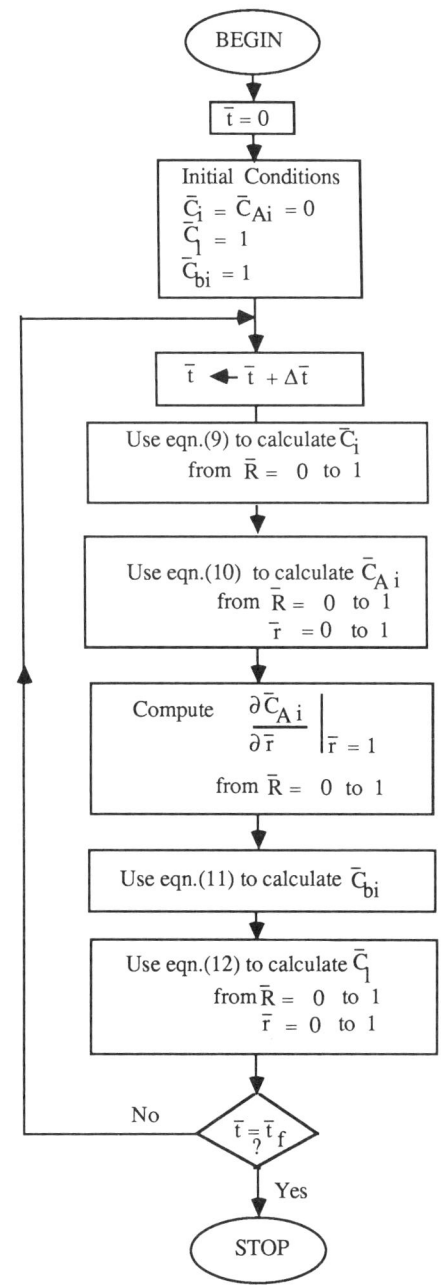

Figure 2. Flowsheet of basic steps in the numerical solution of model equations.

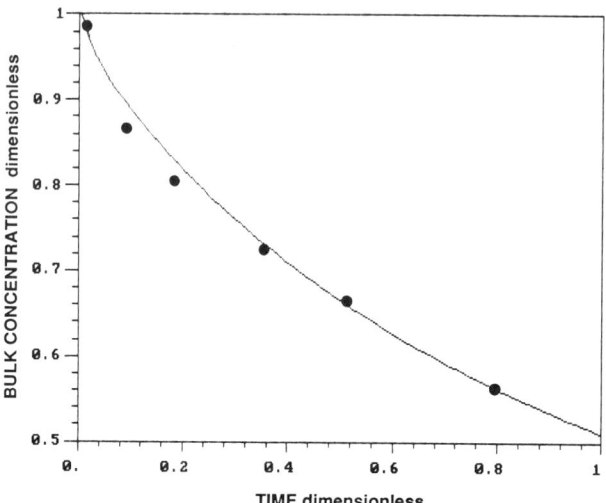

Figure 3. Concentration profile of cycloheximide in a batch adsorber employing immobilized adsorbent beads (see Table I for experimental conditions).

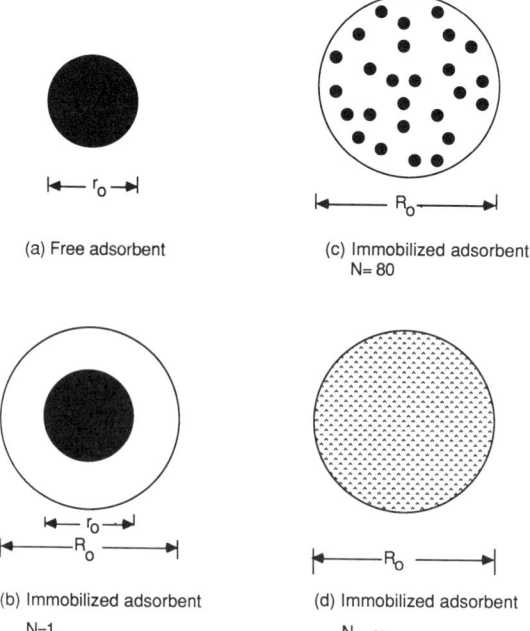

Figure 4. Diagrammatic representation of four cases employed in the simulation studies.

Table I. Physical Parameters used for Simulation Studies

Adsorber parameters:

$V = 50$ ml

$N = 81$

$R_o = R_{ref} = 2.8$ mm

$\varepsilon_a = 0.513$

$n = 107$

$r_o = 0.25$ mm

$C_{bi}^o = 1.0$ gm/liter

$\varepsilon_g = 0.95$

Diffusion and Reaction parameters:

$K_i = 7.05 \times 10^{-3} \text{sec}^{-1}$

$D_i = D_{ref} = 5.8 \times 10^{-6} \text{cm}^2/\text{sec}$

$D_{Ai} = 1.1 \times 10^{-6} \text{cm}^2/\text{sec}$

$\alpha = 0.13$ gm cycloheximide/gm adsorbent

---

equivalent to dispersing the ligand in the hydrogel without another immobilization matrix.

Figure 5 shows numerically generated plots of adsorption rate as a function of time for the above mentioned cases. The adsorption rate was defined as the amount of ligand consumed per unit time using dimensionless units. As expected, addition of the hydrogel layer on the freely suspended adsorbent particle in case (b) causes the mass transfer resistance to go up which reduces the adsorption rate compared to case (a). As shown in Figure 4, the internal mass transfer resistance in (c) is reduced because the adsorbent particles are smaller. This decrease in mass transfer resistance more than overcomes the effect of additional hydrogel resistance. The adsorption rate for (c) therefore shows a sharp increase over that for freely suspended adsorbent particles. This illustrates one of the advantages of using immobilized adsorbent beads over that of freely suspended adsorbent particles. After crushing the adsorbent into an infinite number of particles and dispersing it within the hydrogel bead (case d), only a marginal increase in the adsorption rate over case (c) is observed. This happens because below a certain size the internal mass transfer resistance within the adsorbent particle becomes low enough that it does not control the overall adsorption rate. Based on these results it can be concluded that the adsorption rate increases monotonically with reduction in adsorbent particle size within the hydrogel bead. However, below a certain size the adsorption rate does not increase appreciably. As discussed earlier, there may be added difficulties in recovering very fine adsorbent particles from the bead after dissolving the hydrogel. Thus, optimization of the adsorbent particle size should take into account the additional cost associated with the loss of adsorbent during recovery compared to the advantages of increasing the adsorption rates.

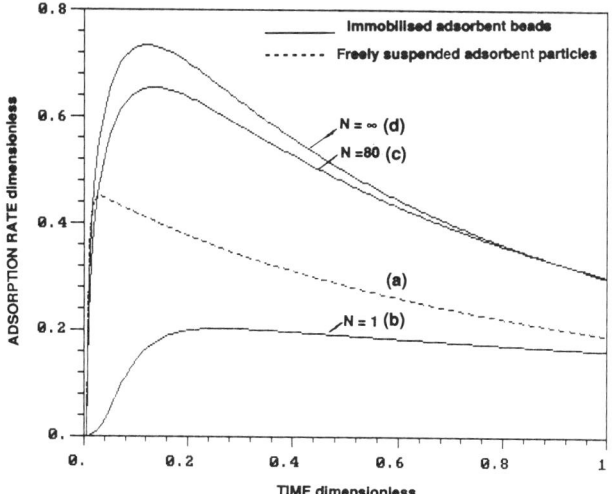

Figure 5. Adsorption rate as a function of time for four simulated cases.

The diffusivity of the desired product in the hydrogel will depend on the gel material, the gel concentration and the degree of cross-linking by multivalent cations. The diffusivity of the product in the adsorbent particles can also vary depending on the choice of the adsorbent matrix used for ligand immobilization. The choice of the hydrogel and the adsorbent matrix will usually depend on several factors such as the stability of the bead against shear forces, the susceptibility to fouling by various agents, and the presence of competing by-products. For efficient bead design one will therefore need to know the effect of diffusivity on product adsorption. Figures 6a and 6b show the effect of varying the product diffusivity in the hydrogel and in the adsorbent matrix respectively. It was found that in both cases, ligands are consumed faster as diffusivities are increased. However, similar to earlier runs the ligand consumption profile approaches a limit as the respective diffusional resistances become smaller.

<u>Two component diffusion and binding</u>. There is a frequent possibility of having one or more compounds present in the fermentation broth which may compete for the available ligands in the adsorbent particles. The objective here is to optimize the bead design so as to maximize the purity of the desired product adsorbed onto the adsorbent particles. In order to numerically simulate such a situation it was assumed that two compounds are being adsorbed onto the immobilized adsorbents: a desired product '1' and an undesired by-product '2'. The adsorption rate constant for the desired product, K_1, is assumed to be 10 times that of the undesired product, K_2. The diffusivities for both of these products are assumed to be similar. Two additional parameters are defined to study the dynamic behavior of such systems.

$$\text{Selectivity (S)} = \frac{\text{Adsorption rate of desired product}}{\text{Adsorption rate of undesired product}}$$

$$= \frac{\sum K_1 C_{A1} C_1 \Delta V}{\sum K_2 C_{A2} C_1 \Delta V} \tag{13}$$

Product purity (Pu) =

$$\frac{\text{Amount of product '1' adsorbed}}{\text{Amount of product '1' adsorbed} + \text{Amount of product '2' adsorbed}} \tag{14}$$

Figure 7 shows the variation of selectivity with respect to time for three types of affinity beads (Cases (a), (b) and (c)). In all three cases, selectivity decreases from the initial maximum value as time progresses. Due to identical diffusivities, the two products have very similar concentration profiles within the immobilized adsorbent bead at initial time. Thus the initial selectivity is just the ratio of their adsorption rate constants. However, since product

(a)

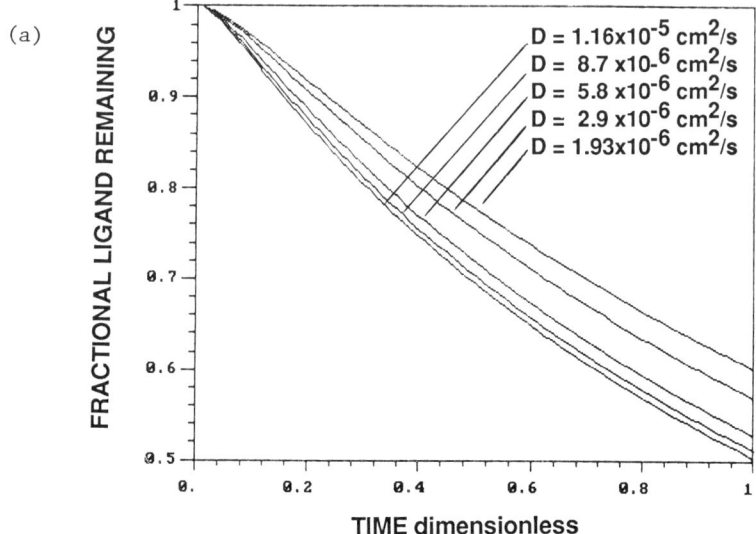

(b)

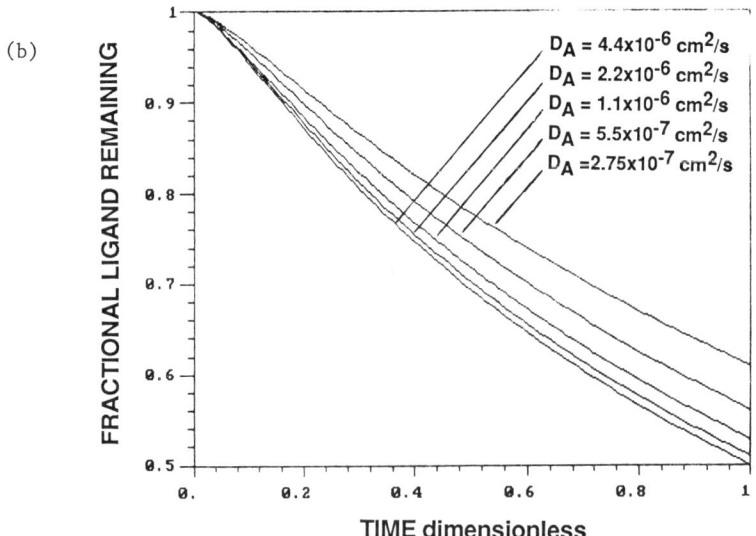

Figures 6a,b. Effect of bioproduct diffusivity in hydrogel (D) and in adsorbent matrix (D_A) on ligand consumption using immobilized adsorbent beads.

'1' is adsorbed at a higher rate, (Figure 7, right), the concentration of product '1' within the bead gradually becomes lower than that of product '2' due to significant diffusional resistance. The bulk concentration of desired product '1' also declines faster than that of the undesired product. The combined effect of these two mechanisms leads to the initial decrease of the selectivity in all three cases. Diffusional resistance effects diminish as the ligand gets consumed and the concentration within the bead becomes closer to the bulk concentration. In some cases, this leads to an increase in the selectivity near the end of the adsorption process.
It was found that the decline in selectivity was least in case (c) because of a smaller overall diffusional resistance of the bead. Figure 8 shows the variation of product purity (Pu) as a function of time for these three cases. The product purity curves show the same general trend as the selectivity curves. Final product purity was also found to be highest for case (c). By virtue of their lower overall mass transfer resistance case (c) immobilized adsorbent beads not only display a higher adsorption rate but also offer a higher selectivity for the desired product.

Conclusions

The use of small adsorbent particles immobilized in hydrogel beads for whole broth processing represents a novel approach to increase the overall extraction yield of biosynthetically derived products. Immobilized adsorbent beads display major advantages over freely suspended adsorbents both in terms of adsorption rate and selectivity. Other practical advantages of these immobilized adsorbent beads are easy handling and reduced fouling characteristics. A mathematical model was developed and used to investigate simultaneous mass transfer and binding within the immobilized adsorbent beads. Numerical simulation of a batch adsorption process employing these immobilized beads was found to be a useful way to study their dynamic behavior and optimal design.

Acknowledgments

We would like to acknowledge the financial support from National Science Foundation which made this work possible.

Legend of Symbols

C_{Ai} product concentration in adsorbent particle, gm/ml

C_i product concentration in hydrogel, gm/ml

C_l ligand concentration (fraction of original binding sites remaining)

C_{lo} initial ligand concentration (1.0)

C_{bi} bulk concentration of the product, gm/ml

C_{bi}^o initial bulk concentration of the product, gm/ml

Legend of Symbols continued on p. 167

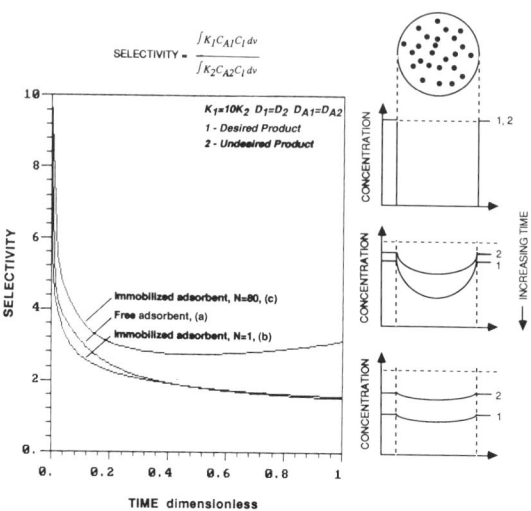

Figure 7. (left) Selectivity as a function of time for competitive adsorption of two compounds. (right) Concentration profiles within the immobilized adsorbent bead and the bulk solution.

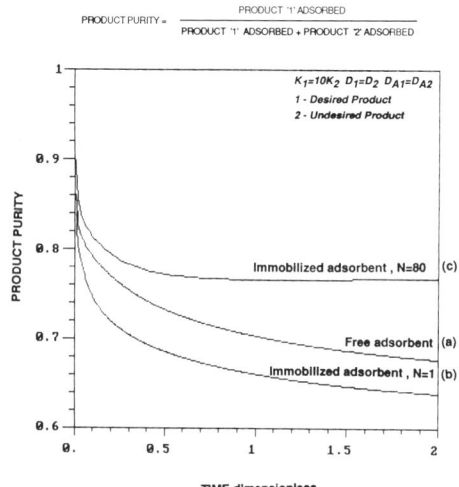

Figure 8. Product purity as a function of time for competitive adsorption of two compounds.

C_{A1}	concentration of desired product in adsorbent particle, gm/ml
C_{A2}	concentration of undesired product in adsorbent particle, gm/ml
r	radial distance within adsorbent particle, cm
r_o	radius of adsorbent particles, cm
R	radial distance in hydrogel bead, cm
R_o	radius of hydrogel bead, cm
R_{ref}	arbitrary reference distance for making the time scale dimensionless, cm
t	time, sec
D_{Ai}	product diffusivity in adsorbent matrix, cm^2/sec
D_i	product diffusivity in hydrogel, cm^2/sec
D_{ref}	arbitrary reference diffusivity for making the time scale dimensionless, cm^2/sec
K_i	adsorption rate constant, 1/sec
N	number of adsorbent particles immobilized within a hydrogel bead
n	number of beads in a batch
NC	number of adsorbing components in the broth
ε_a	porosity of adsorbent matrix
ε_g	porosity of hydrogel
ΔV	volume element inside adsorbent particle
α	ultimate loading capacity, gm/unit ligand

Subscripts:

i   represents i'th adsorbing component in the broth

Superscripts:

—   represents variable in dimensionless form.

Literature Cited

1. Wang, H. Y. Annals of the New York Academy of Sciences, Biochemical Engineering III, 1984, 413, 313.
2. Graves, D. J.; Wu, Y. T. Methods Enzymol. 1974, 34, 140.
3. Chase, H. A. Chem. Eng. Sci., 1984, 39, 1099.
4. Tanaka, H.; Matsumura, M.; Veliky, I. A. Biotech. Bioeng., 1984, 26, 053.
5. Ors, N.; Dogu, R. AIChE J., 1979, 25, 723.

6. Kulkarni, B.D.; Jayaraman, V. K.; Doraiswamy, L. K. Chem. Eng. Sci., 1981, 36, 943.
7. Maheshwari, J.; Nigam, S. C.; Kunzru, D. AIChE J., 1985, 31, 1393.
8. Carnahan, B.; Luther, H. A.; Wilkes, J. O. 'Applied Numerical Methods'; John Wiley Sons; New York, NY, 1969.
9. von Rosenberg, D. U. 'Methods for the Numerical Solution of Partial Differential Equations'; American Elsevier Publishing Co., Inc.; New York, 1969.
10. Wang, H. Y.; Sobnosky, K., unpublished data, 1984.
11. Payne, G. F., Ph.D. Thesis, The University of Michigan, Michigan, 1984.

RECEIVED April 1, 1986

12

High-Resolution, High-Yield Continuous-Flow Electrophoresis

William A. Gobie and Cornelius F. Ivory

Department of Chemical Engineering, University of Notre Dame, Notre Dame, IN 46556

> Recycling effluent through a thin-film continuous flow electrophoresis (CFE) chamber allows virtually complete separation of a binary feed with negligible dilution of products and permits throughput to be increased by O(100-10,000) over present thin-film CFE devices. An approximate model of recycle CFE is developed for the high Peclet number regime and solved analytically. The solution is used to characterize the behavior of a recycle CFE device.

Virtually all biologically derived materials including proteins, nucleic acids and cells, can be characterized by one or more of the popular bench-top electrophoretic techniques. Often the suspending medium resides in an inert matrix which acts not only to stabilize the solvent against natural convection but may also contribute considerably to molecular classification through gel 'sieving' or filtration effects. In free solution electrophoresis retains the ability to resolve homologous species and, because the stresses associated with electrophoretic processing are relatively small, biological activity losses due to mechanical denaturation are generally negligible. However, in free solution one must be extraordinarily careful to avoid excessive thermal excursions which may drive natural convection and/or inactivate product.
    The first free-flow electrophoresis device (1) was designed specifically for large scale biologicals purifications and, although that unit never became fully operational, it preceded a plethora of innovative instruments which either resolved or eliminated the drawbacks inherent in the earliest device. For instance, one apparatus which has been commercially available since the early sixties, the 'thin-film' device (2) is successfully operated at both analytical and preparative scales, e.g.

up to 5ml/hr of a 2% protein solution or about 0.1gm/hr. Further
scale-up of this apparatus is limited by buoyancy-induced convection in the fluid film (3,4) which greatly reduces solute resolution (7).
Two alternative designs have succeeded in stabilizing the
hydrodynamic flow and increasing device capacity by, in one case,
applying a shear stress across a cylindrical annulus with a
radial electric field and axial flow (6,7) and, in the second
case, by operating a 'thin-film' CFE in the low gravity environment available in space (8). The former apparatus is capable of
processing protein solutions at throughputs greater than
2000ml/hr but with significantly poorer resolution than that
obtainable in the thin-film design. The latter column also
achieves throughputs of this magnitude but space shuttle payload
expenses, nearly $100/ounce, render extraterrestrial processing
economically prohibitive in all but a handful of exceptional
cases.

A modified version of the 'thin-film' chamber has been proposed (9) which is capable of processing labile biomaterials at
elevated throughputs while recovering product at arbitrarily high
purity. The proposed design is based on the 'thin-film' device
but uses effluent recycle with prescribed backshift to achieve
effective countercurrency in the electrophoresis column, thus
allowing virtually complete separation of feed into two product
streams. In addition, repeated contact between solute and
electric field allows the power input to be reduced and the
transverse thickness of the chamber to be increased, yielding a
proportionate increase in the device capacity. In the following
paper the performance and characteristics of the modified
electrophoresis chamber are investigated in the limit of
vanishingly small diffusion coefficients by using an approximate
mathematical model.

Continuous Flow Electrophoresis

The classic thin-film CFE device depicted in Figure 1 consists of
a thin, broad chamber through which a laminar curtain of buffered
fluid flows axially. Feedstock is continuously injected into
this flow near the chamber entrance and an electric field imposed
across the curtain causes charged species to migrate laterally as
they pass through the chamber. The solutes form discrete bands
according to their electrophoretic mobilities and elute at the
bottom of the chamber where the individual bands are collected
through a multiplicity of ports.

Joule heat generated by the ionic current is removed through
cooling jackets mounted on the transverse walls. This creates a
transverse temperature gradient which drives a stable and usually
deleterious convective flow (3). Furthermore, the axial temperature gradient in the thermal entrance region (10,11) near the
tips of the electrodes may drive a buoyancy instability when the
thermal gradient exceeds a critical value (4,12). Apart from the
physical properties of the carrier fluid, the critical temperature

gradient required to drive this instability depends on the rate of power dissipation in the fluid and increases with the fifth power of the transverse chamber thickness. Noting that the power input is fixed by the requirement that separation be accomplished in a chamber of moderate length, this strong dependence on the transverse thickness ensures that terrestrial CFEs are generally restricted to chamber thicknesses of O(2mm). Thus the requirement of stable, laminar flow directly limits throughput by constraining the transverse thickness of the device.

The lateral deflection experienced by a particle passing through the chamber depends strongly upon its transverse position since the axial velocity profile is parabolic. Particles nearer the transverse walls are exposed to the electric field longer than those near the centerline, and, in the absence of electroosmosis, migrate a greater lateral distance across the chamber, distorting the solute into crescent shaped bands. Electroosmosis, which is the electrically driven flow of fluid in the double layer adjacent to the charged transverse walls, sets up a lateral circulation pattern, $0_{os}(y)$, which flows parallel to the walls, reverses direction near the electrodes and returns along the chamber centerline (13). The resulting parabolic flow strongly affects solute dispersion by altering the shape of the 'crescent' and, to some extent, the crescent phenomenon may be used to focus solutes into compact bands (14,15) by proper adjustment of the ζ-potential of the transverse surfaces.

Bands distorted by crescent formation will tend to nest inside one another causing the solute elution profiles to overlap. As a result, device resolution suffers and the yield of purified product drops. To reduce crescent formation in the thin-film CFE, the solute feed is usually restricted to the portion of the flow field with the least variation in axial and lateral velocities. This is accomplished by centering the feed stream between the transverse walls and limiting its diameter to no more than 30% of the chamber thickness (2). Therefore crescent formation further reduces throughput by limiting the portion of the chamber through which solute may pass. Note that if the feed were injected through a square port spanning the transverse axis of the chamber, an order of magnitude increase in throughput would immediately be realized. Eliminating or compensating for the dispersive influences associated with crescent formation would therefore yield a significant increase in scale of the 'thin-film' apparatus.

Recycle CFE (RCFE)

In the RCFE effluent is continuously reinjected into the chamber via recycle streams as indicated in Figure 2. Each recycle stream reinjection port is offset from its corresponding elution port by a specified lateral distance, S, so that upon recycle the effluent is shifted back against the solute's electrophoretic migration. When the shift is small solute migrates in the positive z direction but if the shift is increased sufficiently

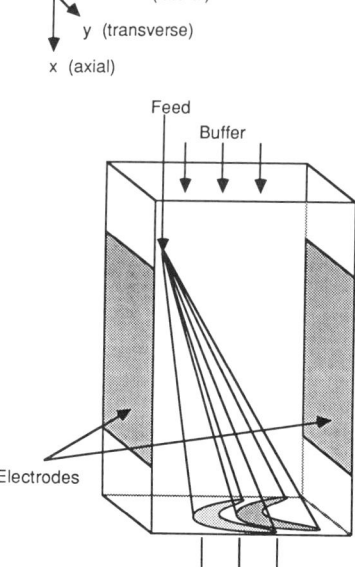

Figure 1. Schematic of the classic CFE device. Solutes injected near the entrance separate laterally as they pass through the chamber. As depicted here crescent formation causes the solutes' elution distributions to overlap.

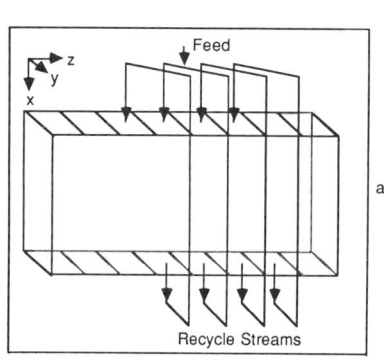 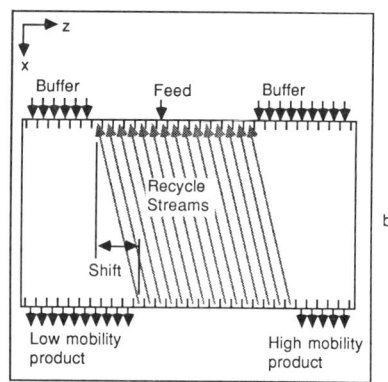

Figure 2. Recycle CFE device. a. Schematic of elution port to inlet port connection. Eluant is continuously reinjected at the chamber inlet. Feed is injected into one of the recycle streams. b. Schematic of complete RCFE device. Recycle streams connect the elution and inlet ports in the recycle section. Separated products are recovered through the elution ports flanking the recycle section, and an equal volume of buffer is fed through the inlet ports on either side of the recycle section.

the solute will begin to migrate in the opposite direction. The value of the shift at which a solute begins to migrate in the opposite direction is referred to below as its 'flip' point. By adjusting either the shift or the electric field strength, two solutes of differing electrophoretic mobility can be made to migrate in opposite directions.

Crescent formation occurs in the recycle chamber to the same degree as is found in the conventional CFE. However, since the effluent includes solute gathered from the transverse wall region as well as from the centerline of the chamber and because the recycle streams are thoroughly mixed before reinjection, after a sufficient number of cycles through the RCFE each solute will have sampled all possible transverse positions and will thus be displaced entirely to the left or to the right side of the chamber. This ensures that the dispersion associated with crescent formation is eventually counteracted by repeated contact with the electric field.

As noted above, when two components with different mobilities are introduced into the chamber, the shift may be adjusted so that the solutes separate in the recycle section of the chamber and elute on either side.

RCFE Model

A model of the RCFE has been developed to help characterize device performance under various operating conditions. To facilitate solution various simplifications have been introduced into the model to allow analytical solution of the mass conservation equation while retaining the essential operating features of the device. For example, in the model presented below the recycle section extends infinitely along the lateral axis and is assumed to have infinitesimally resolved recycle ports so that the governing equation and boundary conditions are continuous. Furthermore, the feed is modeled as a line source of unit strength, extending across the transverse dimension of the chamber at the inlet and no provision is made for product removal at the left and right-hand sides of the chamber.

Solute transport in the interior of the electrophoresis chamber is described by the convective-diffusion equation,

$$\nabla \cdot \{-D\nabla C + \underline{U}C\} = 0 \qquad (1)$$

where D is the molecular diffusion coefficient and $\underline{U} \equiv U_x(y)\underline{i} + O_{os}(y)\underline{k} + \mu E\underline{k}$, including convective, electroosmotic and electrophoretic velocity components. The recycle boundary condition equates the solute flux at the chamber inlet at $x=0$ to the flux through the chamber outlet at $x=L$ while taking into account the lateral backshift of effluent, S, during recycle. Feed injection is included in the boundary condition as an impulse of strength, J_0, located at $(x,z)=(0,0)$.

$$J_x(0,y,z) = \frac{U_x(y)}{2B<U>} \int_{-B}^{B} J_x(L,y,z-S)dy + J_0\delta(y,z) \qquad (2)$$

In the low Peclet number limit, $Pe \equiv <U_x>B/D \ll 1$, an analytical solution of Equation 1 has been derived (16,9) and used to investigate the performance characteristics of the recycle chamber. Similarly, for a packed column the diffusion coefficients are replaced with separate axial and lateral dispersion coefficients computed using Taylor-Aris theory. The resulting equation is linear with constant coefficients and is also amenable to analytical solution. However, in the high Peclet number regime, the parabolic dependence on the transverse coordinate of the convective and electroosmotic velocities severely complicates solution of the governing equations. In general, reported diffusivities of oligopeptides are less than or approximately equal to 10^{-6}cm^2/sec so that Peclet numbers, $Pe \equiv <U_x>B/D$, in the CFE chamber are greater than about 10,000. Because of this diffusion contributes insignificantly to the dispersion and this suggests that a convective dispersion model could be effectively used to provide a good qualitative description of column performance.

To obtain a model of practical utility which still emphasizes the important characteristics of the RCFE in the high Pe limit, it is desirable to substitute a dispersion equation involving the transverse average concentration for Equation 1. The y-dependent velocities are replaced by their transverse averages and the convective dispersion of solute associated with the crescent phenomenon is lumped into a lateral dispersion coefficient, K, which simplifies the analysis considerably and allows analytical solution of the governing equation,

$$<U_x(y)> \frac{\partial <C>}{\partial x} + \mu E \frac{\partial <C>}{\partial z} = K \frac{\partial^2 <C>}{\partial z^2} \qquad (3)$$

This model of convective dispersion in the chamber yields only a fair approximation of the concentration profile for a solute which makes a single pass through the chamber (Figure 3) and, in particular, it neglects the asymmetry in the profile caused by the crescent phenomenon. However, using moments to compute the dispersion coefficient directly from the analytical solution for the flux in a single pass through the chamber allows the approximate dispersion model to qualitatively predict the most important features of the RCFE as will be discussed below. In order that the dispersion model mimic the convective model, K is adjusted so that the solute distribution predicted by each model has identical first and second moments. This ensures that the simplified dispersion model will yield a reasonable estimate of the band spreading.

Figure 4, which is a plot of the dispersion coefficient as a function of the electrophoretic mobility for fixed electric field

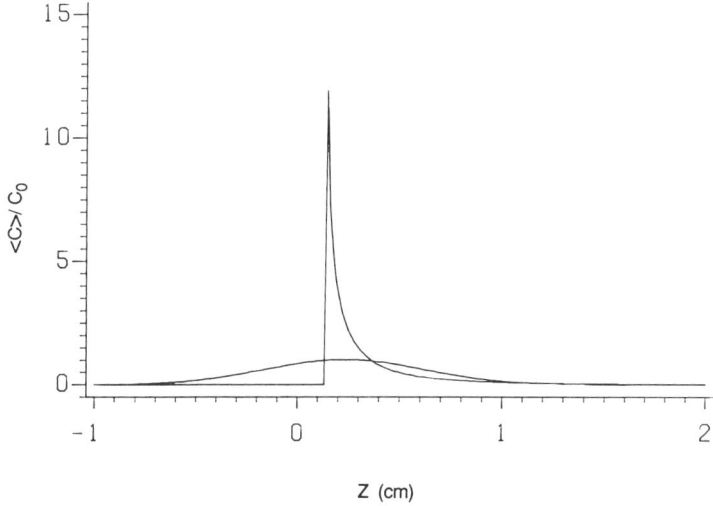

Figure 3. Comparison of single-pass CFE models using parameter values given in Table II. The long tail exhibited by the zero diffusion model (sharp peak) is caused by crescent formation. The dispersion caused by crescent formation has been approximated in the dispersion model (normal curve) by adjusting K so that both distributions have identical variance.

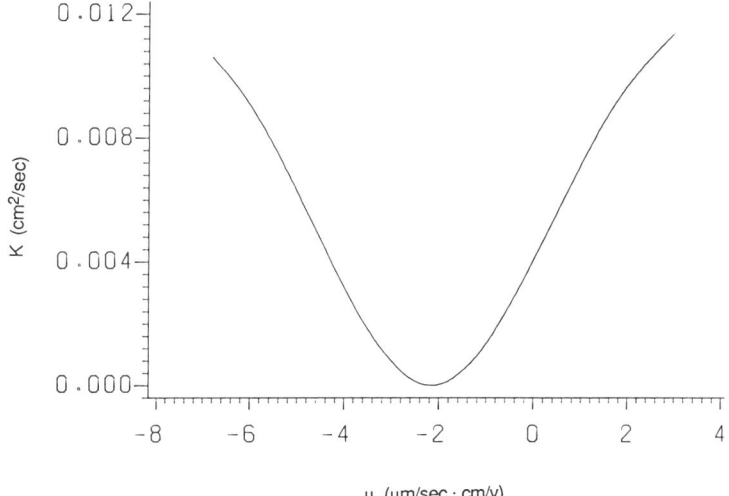

Figure 4. Effective dispersion coefficient, K, vs. electrophoretic mobility, μ. As μ approaches μ_{os}, where μ_{os} = 2.15 μm-cm/v-sec, dispersion caused by crescent formation vanishes. See Table II for parameters used other than indicated in the figure.

strength and electroosmotic wall velocity, demonstrates that a dispersion coefficient computed in this fashion will correctly predict focusing of the solute band (14,15), i.e. a minimum in the dispersion, when the electrophoretic mobility is equal and opposite to the electoosmotic wall mobility. Note that the effective lateral dispersion coefficient, K, is three or more orders of magnitude greater than the molecular diffusion coefficient, $D \approx 10^{-6} cm^2/s$, so diffusion may safely be neglected in this model.

We do not mean to imply that Equation 3 is appropriate as a model only because it can be solved analytically. But in fact, under certain conditions it is possible to obtain anaytical solutions to the first order PDE which results when the diffusive terms are neglected in Equation 1. Under these condtions this model predicts single pass concentration profiles accurately and, when applied to the RCFE it yields values for the 'flip' point, i.e. $S=-L\{\mu E/<U_x>\}$, which are identical to those predicted in the low Peclet number limit, suggesting that this condition is independent of both molecular diffusion and the electroosmotic flowrate for all values of the Peclet number. It is therefore expected that Equation 3 will yield accurate predictions of both the flip point and the flux profiles in the RCFE.

Since this model neglects axial dispersion and uses a transverse average concentration, the recycle condition, Equation 2, reduces to

$$<C(0,z)> = <C_L, z - S> + J_0 \delta(z). \quad (4)$$

The two-sided Laplace transform (17) in z

$$\bar{f}(p) = \int_{-\infty}^{\infty} f(z) e^{-pz} dz \quad (5)$$

is used to obtain the solution to Equation 3 since the shift rule will readily handle the recycle backshift in boundary condition 4. The model equation is solved in transform space and inverted via expansion in residues so the solution is expressed as a series of exponential functions.

Table I. Dimensionless Parameters.

ξ	= xDi/B	ζ =	zDi/B
λ	= LDi/B	σ =	SDi/B
ϵ	= $\mu E/U_x$	Di =	$K/B\, U_x$

The concentration distribution at the chamber exit, in terms of the dimensionless parameters defined in Table I, is determined to be

$$\langle C(\xi,\zeta) \rangle = \text{sgn}(\zeta) \exp(\varepsilon\zeta/2Di^2) \sum_{n=0}^{\infty} \exp(\xi[Di^2 r_n^2 - \varepsilon^2/4Di^2]$$

$$+ r_n\zeta)/\{R'(r_n)\} \quad (6)$$

where the characteristic equation is

$$R(r_n) \equiv 1 - \exp(\lambda[Di^2 r_n^2 - \varepsilon^2/4Di^2] - \sigma[\varepsilon/2Di^2 + r_n]) = 0 \quad (7)$$

with two real zeros,

$$-\frac{\varepsilon}{2Di^2}, \quad \frac{2\sigma + \lambda\varepsilon}{2\lambda Di^2} \quad (8)$$

of which the first can be used to determine the far-field flux and the second used to determine the lengthscale over which the lateral concentration gradient decays to zero. In addition there is a multitude of complex zeros of Equation 7 which moderate solute behavior closer to z=0.

The term in the summation in Equation 6 which involves the first real zero has no z dependence and represents the solute concentration at infinite lateral distance from the feed. This is termed the "far-field" flux in this paper and, in terms of dimensional quantities, it is

$$J^{ff}(z) = \frac{B/L \ \text{sgn} \ (z)}{\left(\frac{\mu E}{U_x} + \frac{S}{L}\right)}$$

The denominator represents the mean displacement of the solute per cycle and its sign indicates the lateral direction, either to the left (-) or right (+), to which the solute migrates under the combined influence of electrophoresis and recycle. In all cases the concentration at the other end of the chamber falls to zero. Note that the far field concentration may become arbitrarily large when the two terms are equal in magnitude and opposite in sign and this is the point at which the far-field flux 'flips' to the opposite side of the chamber.

Table II.
Nominal Values of Parameters Used in Sample Calculations

$2B = 0.375$ cm	$L = 16.0$ cm
$S = 0.0$ cm	$U_x = 0.70$ cm/sec
$\mu = 2.5$ μm-cm/v-sec	$\mu_{os} = +2.15$ μm-cm/v-sec
$D = 1.0 \times 10^{-6}$ cm^2/sec	$K = 1.68 \times 10^{-3}$ cm^2/sec

Discussion

To assess the performance of the RCFE at elevated Peclet numbers several sample calculations were performed using the parameters given in Table II. Recycling allows the electric field strength to be reduced from 70 v/cm in a single pass CFE to 41.25 v/cm in this example and, as a consequence of the reduction in Joule heating, the chamber transverse thickness can be increased by a factor of 1.70. Assuming square inlet ports, the throughput can be increased by a factor of 2.88 over the single pass, 'thin-film' chamber.

The lateral width of the chamber is dictated by the smallest positive and negative (nonzero) arguments of the the exponential functions in Equation 6 since these terms determine the lengthscale over which the concentration gradients decay to zero. In the neighborhood of the feedport Equation 6 converges slowly and dashed lines have been used in Figures 5-7 to indicate approximate concentrations in those regions where adequate convergence could not be obtained even with the use of numerical acceleration techniques (18).

Figure 5 illustrates the effect on a single solute of varying the shift, S, while holding all other parameters constant. When S=0 the effluent is recycled directly overhead and is displaced to the right on subsequent cycles through the chamber. As the magnitude of the backshift is increased, S<0, the relaxation lengthscale increases slowly while the far-field concentration increases rapidly according to Equation 9. When the shift reaches the 'flip' point, the far-field concentration is predicted to be infinite and to fill both sides of the column. If the shift is further perturbed, solute is displaced to the left-hand side of the chamber and, as the magnitude of the backshift is increased still further, the far-field flux decreases but solute elutes on the left-hand side of the chamber.

In Figure 6 the lateral dispersion coefficient, K, is varied by an order of magnitude about the nominal value of 1.7×10^{-3} cm^2/s obtained using the parameters in Table II. For small K an oscillation in the flux appears near z=0 because the feedband, modelled in our calculations as a Dirac distribution, is not entirely dispersed until it has passed through the chamber several times. The three peaks evident in this curve represent the solute on its first, second and third cycles through the chamber, before the impulse has spread out over the lateral axis.

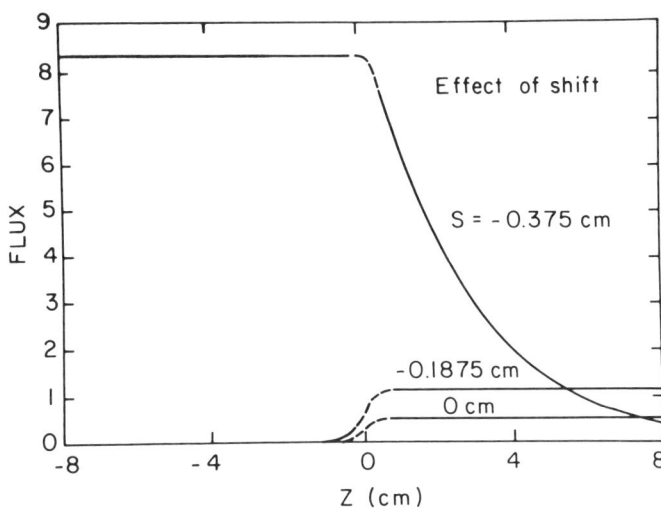

Figure 5. Effect of varying the shift. As the backshift is increased the far-field flux increases as does the length of the toe of the distribution. Once the shift passes the "flip point", the solute migrates opposite to the direction of its electrophoretic motion. See Table II for parameters used other than indicated in the figure.

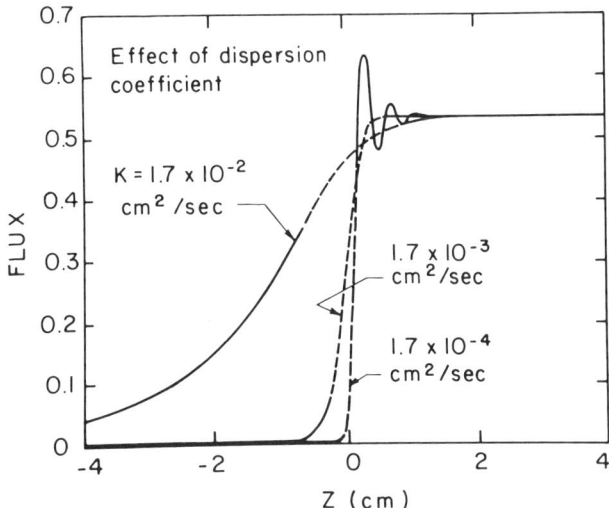

Figure 6. Effect of varying the lateral dispersion coefficient. Dispersion has a marked effect on the relaxation lengthscale. See Table II for parameters used other than indicated in the figure.

As the dispersion coefficient is increased this oscillation disappears since the solute is quickly smeared over the entrance region. For large K, the relaxation lengthscale, which is dictated primarily by the second exponent of Equation 8, increases rapidly with increasing dispersion coefficient since that exponent decreases with the square of the dispersion number, Di.

Figure 7 presents the results of a simulated binary separation of solutes which have mobilities of 2.5µ-cm/v-s and 3.0µ-cm/v-s, respectively, with shift, S=-0.375cm. The relative fluxes shown indicate that the low mobility component is concentrated about 6x above its feed value while the faster moving component is concentrated to about 4x its feed value. Note that the separation is effected in a chamber about 16cm wide and that both solutes can be recovered at arbitrarily high purities by extending the breadth of the recycle section.

Table III.
Typical Electrokinetic Parameters of Colloidal and Biological Materials

(µm-cm/volt/sec)		μ_1		μ_2
Serum Proteins:	(Hemoglobin)	0.12	(Albumin)	0.59
Average Axial Velocity: 0.074 cm/sec				
Red Blood Cells:	(Human RBC)	1.16	(Sheep RBC)	1.44
Average Axial Velocity: 0.093 cm/sec				
Polystyrene Latex Particles:	(.2µm dia.)	6.5	(.8µm dia.)	9.2
Average Axial Velocity: 0.235 cm/sec				

Electric Field Strength: 25 volts/cm
Electrode Length: 16 cm
Electrophoretic Wall Mobility: -1.0 µm-cm/volt-sec

To help put these calculations in perspective Table III contains typical values of the electrophoretic mobilities measured for various materials of widely varying size. In general, one should expect lower molecular weight particles to have mobilities in the range, 0.1-3.0 µm-cm/volt-sec, depending on the composition of the carrier fluid while cells and particulates would have mobilities in the range, 0.5-10.0 µm-cm/volt-sec. However, differences in the relative magnitudes of the mobilities between the various groups cited above are readily compensated by adjusting the carrier fluid flow-rate so that pairs of species which have low mobilities are exposed longer to the electric field. Recall that in the hypothetical RCFE modelled in this paper, several simplifying assumptions have been invoked to facilitate investigation of the operating characteristics of the device and one of these assumptions was that no provision be made in the model for product withdrawal. The addition of withdrawal ports at either end of the RCFE will cause the concentrations of product to drop below their feed concentrations, thwarting attempts to concentrate solute. However, this problem is circumvented by adding "regenerators" on either end of the column as described in Figure 8.

The regenerators are essentially additional recycle sections appended to either side of the main recycle section in which the shift is altered by one or more port widths. When the shift is properly adjusted, each solute's net lateral velocity is reversed in the regenerators and upon entering the regenerator solute is forced to return to the recycle section. Since the solute can exit the chamber only through the elution ports located between the recycle section and regenerator, the concentration in the product stream will approach the original concentration in the feed. Further concentration of products using the RCFE with regenerators can be effected by using multiple feeds, decreasing the width of the outlet ports or by recycling a portion of the product stream back into the chamber near the inner edge of the regenerator section.

Conclusion

As has been demonstrated in this paper in the high Peclet number limit, continuous effluent recycle with shift allows binary separation of solutes to arbitrarily high purities with essentially complete recovery of products. In addition, products can be recovered at or above their feed concentrations and, because the electric field can be significantly reduced when solutes are repeatedly cycled through the chamber, the transverse thickness of the chamber can be increased proportionately.

Several factors can be taken advantage of to increase throughput in the RCFE over the thin-film design on which the chamber is based: First, using the entire transverse thickness of the chamber and assuming square inlet ports increases throughput by $O(10)$. Reducing the electric field strength and reorienting the device into a more stable configuration (18) allows a further $O(10-100)$ increase in throughput. A further $O(10)$ increase in throughput is made possible by using multiple feed streams, thus an overall increase in the capacity of thin-film type continuous electrophoretic separations of $O(100-10,000)$ can be achieved by utilizing recycle.

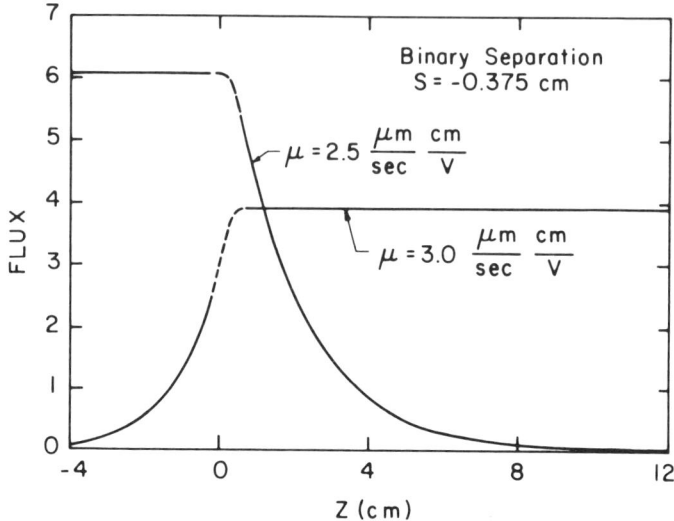

Figure 7. Separation of solutes differing by 20% in mobility. A short distance from the feed point, the purity of both components varies exponentially with z. See Table II for parameters used other than indicated in the figure.

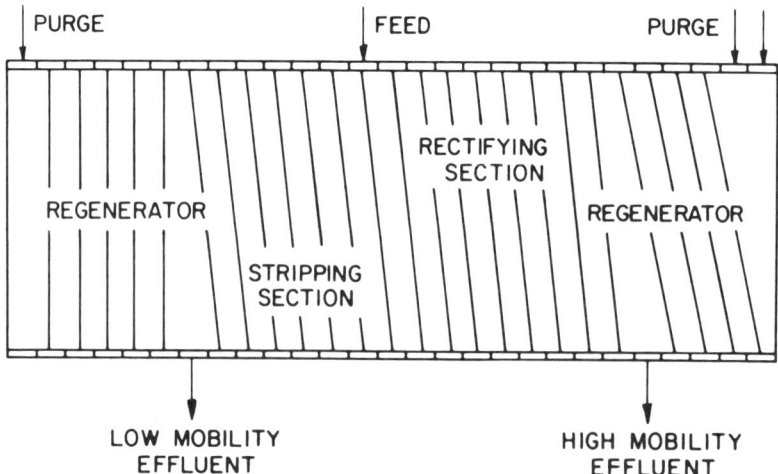

Figure 8. Schematic of RCFE with regenerators. The shift in the additional recycle sections is chosen to reverse the solute's direction of migration, constraining it to exit the chamber through the elution ports between the recycle section and the regenerators. Products are recovered at or near their feed concentration.

Acknowledgment

This material is based upon work supported in part by the National Science Foundation under Grants No. CPE-8211483 and CBT-8414218.

Legend of Symbols

B	-	Half thickness of electrophoresis chamber.
C	-	Concentration.
<C>	-	Average concentration.
D	-	Molecular diffusion coefficient.
D_i	-	Dimensionless dispersion coefficient.
E	-	Electric field strength.
J_0	-	Feed source strength.
J_x	-	Axial flux.
J^{ff}	-	Far-field flux.
K	-	Length of electrophoresis chamber.
L	-	Length of electrophoresis chamber.
O_{os}	-	Electroosmotic velocity profile.
r_n	-	Roots of the characteristic equation.
$R(r_n)$	-	Characteristic equation for r_n.
S	-	Shift distance.
$\underline{U}$	-	Velocity vector.
U_x	-	Axial velocity.
$<U_x>$	-	Average axial velocity.
x,y,z	-	axial, transverse, lateral coordinates.
δ	-	Dirac distribution.
ε	-	Dimensionless electrophoretic velocity.
ζ	-	Dimensionless lateral coordinate.
λ	-	Dimensionless chamber length.
μ	-	Electrophoretic mobility.
μ_{os}	-	Electroosmotic wall mobility.
ξ	-	Dimensionless axial coordinate.
σ	-	Dimensionless shift distance.
∇	-	Gradient operator.

Literature Cited

1. Philpot, J. St. L. Trans. Farad. Soc 1940, 36, 38.
2. Hannig, K. In "Techniques of Biochemical and Biophysical Morphology"; Glick, D.; Rosenbaum, R., Eds.; Wiley: New York, 1972; Vol. 1, pp. 191-232.
3. Ostrach, S. J. Chrom. 1977, 140, 187-195.
4. Saville, D. A.; Ostrach, S. "Fluid Mechanics of Continuous Flow Electrophoresis"; Final Report, Contract NAS-8-31349 Code 361, 1978.
5. Rhodes, P. H. "Sample Stream Distortion Modeled in Continuous Flow Electrophoresis"; NASA TM-78178, 1979.
6. Philpot, J. St. L. In "Methodological Developments in Biochemistry"; Reid, E., Ed.; 1966; Longmans: England; Vol. 2, pp. 81-85.
7. Mattock, P.; Aitchison, G. F.; Thomson, A. R. Sep. Pur. Meth. 1980, 9, 1.
8. Morrison, D. R.; Barlow, G. H.; Cleveland, C.; et al. Adv. Space Res. 1984, 4, 67-76.
9. Gobie, W. A.; Beckwith, J. B.; Ivory, C. F. Biotechnology Prog. 1985, 1, 60.
10. Lynch, E. D.; Saville, D. A. Chem. Eng. Commun. 1981, 9, 201-211.
11. Naumann, R. J.; Rhodes, P. H. Sep. Sci. 1984, 19,51.
12. Saville, D. A. PhysicoChemical Hydro. 1980, 1, 297-307.
13. Nee, T. W. J. Chrom. 1975, 105, 231.
14. Strickler, A. Sep. Sci. 1967, 2, 335.
15. Strickler, A.; Sacks, T. Ann. New York Acad. Sci. 1973, 209, 497.
16. Ivory, C. F.; Gobie, W. A.; Turk, R. S. Electrophoresis '83, 1984, p. 293.
17. van der Pol, B.; H. Bremmer, "Operational Calculus Based on the Two-Sided Laplace Integral"; Cambridge University Press; 1950.
18. Turk, R. S.; Ivory, C. F. Chem. Eng. Sci., 1984, 39, 851.

RECEIVED April 8, 1986

13
Scale-Up of Isoelectric Focusing

Milan Bier

Biophysics Technology Laboratory, University of Arizona, Tucson, AZ 85721

>   The paper describes some applications to large scale protein
> fractionation using a recycling isoelectric focusing apparatus.
> Separation is achieved in free solution without the use of
> supporting media. Various alternatives for the formation of
> the pH gradient are discussed and results of a computer
> simulation are presented.

We are presently witnessing revolutionary developments in applied biology due to the rapid advances in genetic engineering through recombinant DNA and hybridoma technologies. The progress in these areas has surpassed even the most optimistic projections of just a few years ago. The economic impact of these technologies has been amply covered in the scientific and lay press, including reviews by the Office of Technology Assessment of the U.S. Congress (1).

At present, these technologies have rendered possible the production of virtually unlimited quantities of important new biologics which were previously available only in minutest quantities. Human insulin, interferon, human growth hormone, foot and mouth disease vaccine are but a few examples. For the purpose of this symposium it should be emphasized that these proteins are often first obtained in the form of crude extracts, heavily contaminated by extraneous matter, derived from the host organism. The purification of the desired end product is essential if it is to be used as a pharmaceutical. The magnitude of the problem can best be comprehended if one realizes that the host cell, i.e. the modified microorganism or the hybridoma cell, may contain well over 5,000 different proteins, only one of which may be the desired active principle.

Isoelectric Focusing

Isoelectric focusing (IEF) is unique among separation processes as it results in a stationary steady state distribution of fractions along the column axis. The final distribution of fractions is independent of their initial distribution. As such, IEF has no analogue in other electrophoretic or chromatographic methods and well deserves its current popularity (2).

0097-6156/86/0314-0185$06.00/0
© 1986 American Chemical Society

IEF is applicable only to amphoteric compounds, primarily proteins or larger peptides, i.e., compounds which have acidic and basic dissociable groups. As a result, such compounds acquire a positive net charge in acidic media and a negative net charge in basic media. The point of charge inversion, i.e., where the compounds exhibit zero net charge, is refered to as the isoelectric point. When such species are exposed to a d.c. electric field of proper polarity in a pH gradient, they will migrate electrophoretically toward the pH corresponding to their isoelectric point, where they become virtually immobilized as diffusion is balanced by electrophoretic focusing. This final distribution of fractions is independent of their initial distribution within the pH gradient.

Isoelectric point and molecular weight are the two most characteristic parameters of a protein. This is the reason analytical IEF has so rapidly gained widespread usage in research as well as in quality control of product development. The sequential separation according to IEF and molecular weight in two-dimensional electrophoresis gives the sharpest resolution of protein mixtures and permits the recognition of several thousand individual protein species in extracts of various biological tissues (3).

Analytical IEF is routinely carried out in gels of polyacrylamide or agarose, the pH gradient being formed 'naturally', i.e. through the effect of the electrical current itself. Special buffer mixtures were developed for this purpose, the first and best known being available under the tradename 'Ampholine'. It comprises the products of random polymerization of a mixture of polyamines and acrylic acid, thus containing a large number of molecular species. Each species focuses to its isoelectric point and establishes thereby the pH gradient.

Our laboratory has undertaken the task of adapting the principles of IEF to large scale preparative purposes. We have taken a two pronged approach:

1. Development of a better theoretical understanding of all electrophoretic processes, IEF included, through mathematical modeling and computer simulation. Our collaborators, Drs. Saville and Palusinski, have presented this work in a paper (#5e) in the Symposium on Novel Separation Techniques in Biotechnology I at the present AIChE meeting. Some of the results were previously reported (4,5).

2. Development of instrumentation for large scale preparative IEF. Gels are unsuitable for this purpose and separation in free solution is essential. This entails the solution to the two problems common to all electrophoretic processes: stabilization of fluid against convection and dissipation of Joule heat. Our first apparatus, dubbed RIEF (Recycling IEF) solves these two problems in a unique fashion, by using screen elements for streamlining the fluid flow and dissipating the Joule heat in heat-exchangers external to the focusing apparatus (6). The apparatus is of modular design and could be applied, at least in principle, to industrial scale processing. The demand for large scale processing has been lagging, however, and the increased use of the RIEF for research scale separations has prompted us to develop other instruments with smaller throughputs (7).

Recycling Isoelectric Focusing Apparatus

In most electrophoretic methods constituents separate according to differences in their mobility, i.e., the rate of their migration. Scaling to industrially meaningful throughputs is complicated by two problems: the need to dissipate the Joule heat generated by the electric current and the need to stabilize the fluid against convective remixing. IEF, to the contrary, is not based on differences in migration rates, but instead a steady state is obtained, proteins distributing themselves within the pH gradient in accordance to their isoelectric points. We have taken advantage of this unique property of IEF to solve the dual problem of heat dissipation and fluid stabilization by physically separating these two functions.

The apparatus is presented schematically in Figure 1. The solution to be fractionated is continuously recycled between a multichannel focusing apparatus and a multichannel heat-exchange reservoir. Fluid flow through the focusing apparatus is streamlined by means of a parallel array of baffles. Monofilament nylon screens of fine porosity have been found most suitable for this purpose as they allow free transport of all constituents, while still acting as convective barriers. Joule heat is dissipated within the heat-exchange reservoir, which is external to the focusing apparatus itself (6).

Key components are the focusing cell and the heat exchange reservoirs. Fluid is recirculated between the corresponding channels of these two components by the multichannel pump. We have also designed monitors for intermittent registration of protein concentration (through ultraviolet adsorption) and pH of each channel. While not essential for separations, these monitors are under the control of a Hewlett-Packard desk top computer (not shown in the diagram). The computer provides for numerical scaling and calibration of the multiplexed monitors, receives the raw data from the interface at preset time intervals, converts these into optical density and pH units, provides printout of data in real time, stores them on tape, or can display them on a plotter. In addition, it can be programmed for a variety of statistical analyses or feed-back control of the separation process.

Illustrative Fractionation Results

The RIEF system is completely modular and is not restricted to any preset number of compartments, reservoir capacity or cross-area of the focusing cell. Nevertheless, most of our data were obtained with a ten compartment assembly. The partitioning of the focusing apparatus by the screen baffles imposes a step-changing pH gradient, rather than a linear one. The resolution is largely dictated by the pH range of Ampholine used. For highest resolution, very narrow pH range Ampholine is needed. This can be obtained in a preliminary fractionation in the RIEF, followed by the reprocessing of selected fractions in a second run.

Figure 2 illustrates the fractionation of a complex sample, an extract from Bermuda grass, which is a common allergen. Because of the heterogeneity of the sample, a broad range Ampholine was used, pH 3.5 - 10. The results of the fractionation were assessed by analytical polyacrylamide gel isoelectric focusing. Shown are the patterns of the original mixture and of the ten RIEF fractions. As can be

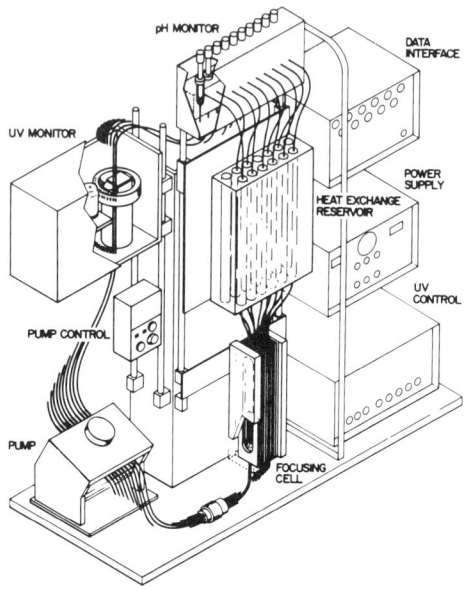

Figure 1. Schematic presentation of the recycling isoelectric focusing apparatus (RIEF). Reproduced with permission from Ref. 6. Copyright 1979, Pierce Chemical Company.

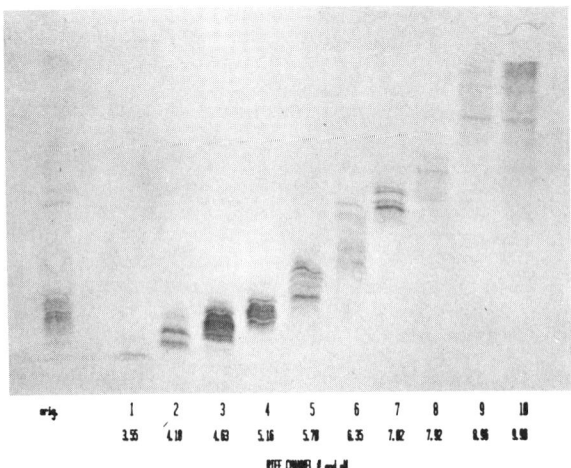

Figure 2. RIEF separation of an aqueous extract of Bermuda grass allergens. Reproduced is a photograph of an analytical polyacrylamide gel focusing pattern of collected samples from the ten RIEF channels.

seen, the RIEF has separated the original mixture into fractions of increasingly basic isoelectric points. The fractions were tested clinically and all found to be similarly allergenic.

In Figure 3 is documented the resolution of a polyclonal antibody sample exhibiting only a few closely spaced isoelectric bands. The antibodies were raised to the bacterial carbohydrates derived from Micrococcus lysodeiktikus in a rabbit from a colony outbred for simplicity of their clonotype patterns. After purification by affinity chromatography, the antibodies exhibited three major bands, with very close isoelectric points. To effectuate their separation, commercial Ampholine was subfractionated in the RIEF and a narrow cut, pH range 7.5 to 8.5, was used for the fractionation (8). This illustrates the resolution achievable in critical separations.

Formation of pH Gradients

Ampholine and other similar carrier ampholytes are generally used for the formation of pH gradients. They contain, however, chemically ill-defined components which may contaminate the purified products. The development of other means for the generation of pH gradients would be highly desirable and was a prime objective of much of our theoretical modeling. Three alternatives were pursued:
1. Focused (static) pH gradients: Mixtures of well characterized low molecular compounds can be used instead of the ill-defined polymeric buffer systems. A stable pH gradient is again generated through the focusing of components to their isoelectric point. In Figure 4 we illustrate this approach with the computer simulation of the focusing of a mixture of cacodylic acid, histidine and tris-(hydroxymethyl)aminomethane (tris), a weak acid, an ampholyte and a weak base, respectively. Starting from a uniform distribution of all three components, the focusing process begins with the accumulation of the acid at the anode and of the base at the cathode. This generates a pH gradient which propagates towards the center of the column, where the amphoteric histidine accumulates. This approach can be utilized experimentally for the focusing of proteins (9), but it suffers from the paucity of suitable compounds. The main amphoteric compounds available are amino acids and oligopeptides, and their isoelectric points do not adequately cover the entire pH range.
2. Dynamically formed pH gradients: Rather than allowing the focusing of components to their isoelectric point, it is possible to generate pH gradients by maintaining a constant flux of buffering electrolytes across the column. This requires the maintenance of constant boundary conditions at the two ends of the column. In principle, this could be accomplished by maintaining large electrolyte volumes at the column ends, with an acidic buffer at the anode and a basic buffer at the cathode. In practice, a reconstitution of the composition of the two buffers seemed more appropriate and various approaches were studied experimentally as well as through computer simulation (10).
3. Preformed pH gradients: It is possible to preform in free solution a pH gradient, using simple buffer components, such as, for example, the above mentioned cacodylic acid and tris. Such gradients are unstable, gradually degrading through ion transport. Nevertheless, we have demonstrated that such gradients are usable in the neutral pH region, about pH 6 to 8, provided only two components are used (11).

190 SEPARATION, RECOVERY, AND PURIFICATION IN BIOTECHNOLOGY

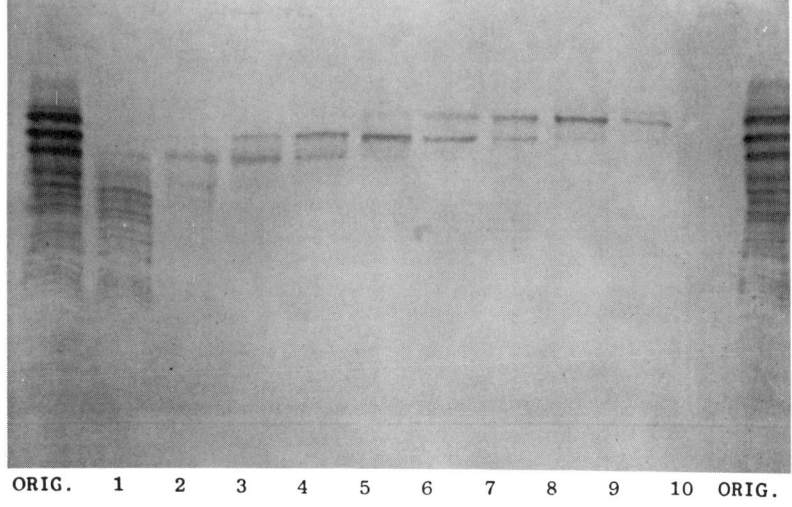

ORIG.　1　2　3　4　5　6　7　8　9　10　ORIG.
pH:　　6.08 7.28 7.49 7.60 7.71 7.81 7.90 8.00 8.21 9.18

Figure 3. Polyacrylamide gel pattern of ten RIEF fractions of a polyclonal rabbit antibody sample.

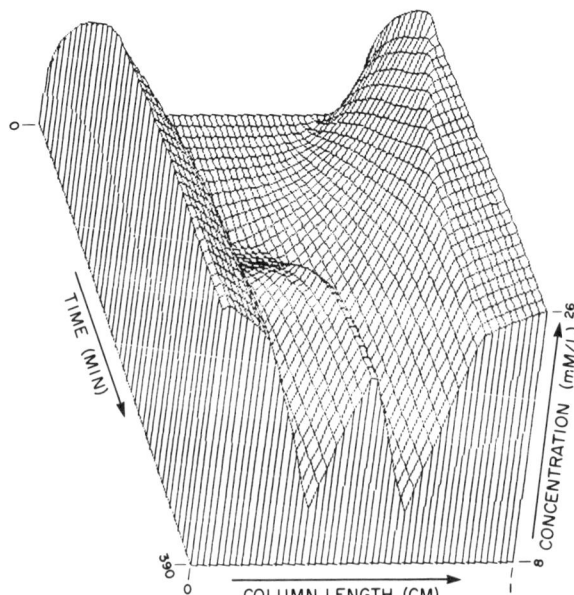

Figure 4. Three dimensional plot of computer simulation data of the focusing process of a mixture of cacodylic acid, histidine and Tris. Reproduced with permission from Ref. 5. Copyright 1983, American Association for the Advancement of Science.

Deterioration of gradients could be avoided by physical immobilization of the functional groups of the buffer. This has been accomplished in analytical gels (12) through the use of Immobilines (Tradename of LKB Produkter, A.B., Bromma, Sweden) but these have not yet found application in free solutions.

Discussion

In principle, the modular design of the RIEF would permit the scaling up of its capacity to industrially meaningful volumes. At present, volumes ranging from 300 to 10,000 ml are being processed, this usually requiring two to four hours. Apparatus cross-sections of 10, 20 and 100 cm^2 are utilized. In other electrochemical instruments of somewhat similar type, such as electrodialysis or forced-flow electrophoresis, much larger cross-sections are utilized. Corresponding extension of the RIEF is quite feasible.

The demand for large scale focusing has been lagging, however, due to several factors: the well entrenched status of chromatography, the lack of off-the-shelf large scale focusing equipment and the need as yet to use Ampholine-like buffers for generation of the pH gradients. Our equipment has found, however, increasing demand for research applications on smaller scale. This has prompted us to design a smaller apparatus, based on a somewhat different principle, with a priming volume of only 40 ml subdivided into 20 fractions (7).

The RIEF system has been utlized for the fractionation of a large number of samples, most of which were provided to us by researchers from industry or academia. These encompassed fermentation products, such as recombinant interferon, products of mammalian tissue culture and a great variety of other proteins, enzymes, synthetic peptide hormones, etc. In general, preparative IEF is particularly well suited for the purification of products of genetic engineering, as they tend to be more homogeneous than natural proteins. This is due to the avoidance in non-mammalian systems of glycosylation, a process secondary to DNA transcription, which accounts for much of the heterogeneity of natural products. The same is true for monoclonal antibodies, which are obviously more homogeneous than the polyclonal ones.

Center for Separation Science

Our laboratory has recently been selected by the National Aeronautics and Space Administration (NASA) as one of its two national centers of excellence in separation science. This has given us an opportunity to broaden our efforts towards the advancement of all modes of electrophoresis for large scale processing. Thus, our Center has been chosen by European manufacturers as the demonstration site for the United States of two unique electrophoretic instruments.

CJB Developments Ltd of Portsmouth, England, has installed at our Center its BIOSTREAM (TM) production scale electrophoresis system. Developed at the Harwell Atomic Energy Laboratory in UK, this apparatus is capable of separating industrially meaningful quantities of proteins, having a throughput of up to 100 grams of protein per hour. Residence time in the apparatus is of the order of a few seconds only, fluid being stabilized against convection through shear induced by an ingenious rotating electrode assembly.

Bender & Hobein Gmbh of Munich, West Germany, has contributed to the Center its ELPHOR Vap 21 free flow electrophoresis apparatus. Of the type flown by McDonnell Douglas Corp. aboard the NASA space shuttle, this apparatus embodies the principles first developed by Hannig (13). It is an innovative apparatus best suited for separation of living cells and cell organelles.

In addition, our Center remains dedicated to the development of novel fractionation methods based on the recycling principle embodied in our RIEF apparatus. Several new prototypes are being tested. Experiments aboard the space shuttle are also planned, as part of our continuing study of the potential advantages in terms of throughput and resolution derivable from operation in the microgravity environment prevailing in orbiting spacecraft.

Acknowledgments

Supported in part by NASA grant NSG-7333 and by NASA contract NAS8-32950.

Literature Cited

1. Commercial Biotechnology: An International Analysis, Office of Technology Assessment, U.S. Congress, Washington, D.C., 1984.
2. P.G. Righetti: Isoelectric Focusing: Theory, Methodology and Applications, Elsevier Biomedical, New York, N.Y. 1983.
3. D.W. Sammons, L.D. Adams and E.E. Nishizawa, Electrophoresis 2, 135, 1981.
4. O.A. Palusinski, T.T. Allgyer, R.A. Mosher, M. Bier and D.A. Saville, Biophys. Chem. 13, 193, 1981.
5. M. Bier, O.A. Palusinski, R.A. Mosher and D.A. Saville, Science 219, 1281, 1983.
6. M. Bier, N.B. Egen, T.T. Allgyer, G.E. Twitty and R.A. Mosher in "Peptides: Structure and Biological Function"; E. Gross and J. Meienhofer, Eds., pp. 35-48, Pierce Chemical Co., Rockford, IL, 1979.
7. N.B. Egen, W. Thormann, G.E. Twitty and M. Bier in "Electrophoresis '83"; H. Hirai, Ed., pp. 547-550, de Gruyter, New York, NY, 1984.
8. S.B. Binion, L.S. Rodkey, N.B. Egen and M. Bier, Electrophoresis 3, 284, 1982.
9. M. Bier, R.A. Mosher and A.O. Palusinski, J. Chromatogr. 211, 313, 1981.
10. A. Tsai: "Study of a Coupled System of Two Electrophoretic Columns with Opposing Current Polarity", Master Thesis, University of Arizona, Tucson, AZ 1984.
11. M. Bier, R.A. Mosher, W. Thormann and A. Graham in "Electrophoresis '83', H. Hirai, Ed., pp. 99-108, de Gruyter, New York, NY, 1984.
12. B. Bjellqvist, K. Ek, P.G. Righetti, E. Gianazza, A. Georg, R. Westermeier and W. Postel, J. Biochem. Biophys. Methods 6, 317, 1982.
13. K. Hannig in "Electrophoresis"; Vol. II, M. Bier, Ed., pp. 423-472, Academic Press, New York, 1967.

RECEIVED March 26, 1986

14

Large-Scale Gel Chromatography
Assessment of Utility for Purification of Protein Products from Microbial Sources

James J. Kelley[1,3], George Y. Wang[1,4], and Henry Y. Wang[2]

[1] Molecular Genetics, Inc., Minnetonka, MN 55343
[2] Department of Chemical Engineering, The University of Michigan, Ann Arbor, MI 48109

> The cost of large-scale gel chromatography fractionation was found to be on the order of $5/gm of product protein, a cost which limits its applicability to high price products, like human pharmaceuticals. The high cost is due to inherent limitations of the gel permeation phenomenon, namely, poor resolution, restricted throughput and product dilution. A quantitative model for assessing the utility of gel chromatography for protein purification was developed. This methodology may be used to determine the ability of gel chromatography to perform a desired separation at the process-scale, and perform design calculations and cost projection analyses.

Since its discovery 25 years ago (1), gel chromatography has been applied extensively at the laboratory scale for analytical and preparative work and industrially for the production of polymers and proteins. Several factors which make it attractive for large-scale production of proteins are:

1.) It is the only method for protein fractionation based solely on molecular size;
2.) Solute concentration is limited only by solubility and viscosity considerations;
3.) It is gentle and thus high recoveries are usually obtained;
4.) Operation is isocratic with respect to pH, buffer composition, and temperature.

[3] Current address: The Stroh Brewery Company, Detroit, MI 48207
[4] Current address: Amoco Research Center, Standard Oil (Indiana), Naperville, IL 60566

The use of gel chromatography has remained high over the years. We surveyed the 1982 volume of The Journal of Biological Chemistry and found that 60% of the published laboratory-scale protein purification schemes included at least one gel chromatography step. This is essentially the same result found by Dunnill and Lilly in 1968 (2).

While there are examples of industrial use of gel chromatography for protein purification and even though it continues to enjoy great popularity among biochemists for small-scale separations, we perceive a hesitancy to apply the method within the Biotechnology industry, due to some inherent problems. These problems are:

1.) Low productivities due to limited feed and flow rate capabilities;
2.) Low column efficiencies;
3.) Solute dilution;
4.) Lack of information about costs involved in large-scale operations.

In order to examine the validity of these suspicions, we first calculated the volumetric productivities of industrial and large-scale gel chromatography protein fractionations published in the literature (Table I). Only applications involving column volumes greater than 4 liters were considered. The first seven examples (3-8) used soft, compressible gels, like Sephadex G-200 and Ultrogel AcA34. Productivities varied from 0.0016 to 0.045 liter feed/liter gel/20 h day (l/l/d) and averaged about 0.025 l/l/d. The next three examples (9-11) used less compressible gels, such as Sepharose 4B, Spehadex G-75 and G-50. The productivities were on the order of 0.1 l/l/d, about 4-fold higher than the compressible gels. The productivity of a continuous annular chromatograph (12) was found to be approximately 0.37 l/l/d, 15-fold higher than the compressible gels. Finally, when productivities for hypothetical HPLC production systems were calculated for Bio-Sil 250 (13) and Licrosorb diol (14), they were found to be approximately 25- and 250- fold higher than compressible gels, respectively.

We also estimated the column efficiency from the elution profile for the industrial separation of insulin (10). It was found to be 3-4 fold less than that predicted by the Giddings plate height equation with reasonable assumptions for gel chromatography (15,16). Thus the restricted productivities often associated with gel chromatography are the result of the predominant use of soft, compressible gels and low column efficiencies are probably due to losses outside the columns. Except for one case (10), there was no discussion in any of these publications concerning how scale-up was achieved, how values for design variables were chosen, or even why gel chromatography was used.

Current Design Methods. Two methods have been used for gel chromatography system design (Figure 1). The first method, which we have called conventional scale-up, is that commonly used by biochemists for preparative work. The conditions for an acceptable separation are first worked out in small columns in the laboratory.

TABLE I

Protein	Separation Medium	Bed Dimensions (D x L)	Productivity	Reference
		(cm x cm)	(1 feed/ 1 gel/day)	
E. coli Iso-leucyl-t-RNA transferase	Sephadex G-200	21.5 x 80	0.026	3
E. coli Methionyl-t-RNA transferase	Sephadex G-200	14 x 180	0.0016	4
Citrobacter L-asparaginase	Sephadex G-200	4.2 L	0.007	5
Rape Seed Proteins	Sephadex G-200	45 x 85	0.025	6
Transferrin from Cohn Fraction IV	Sephadex G-200	45 x 75	0.019	7
Plasma Alkaline Phosphatase	Ultrogel AcA 34	16 x 100	0.045	8
Plasma Cholinesterase	Ultrogel AcA 34	9.3 x 90	0.026	8
Thyroglobulins	Sepharose 4B	37 x 45	0.125	9
Insulin	Sephadex G-50	37 x 90	0.071	10
Whey Protein Concentrate	Sephadex G-75	37 x 15	0.128	11
Continuous, annular Chromatograph	Sephadex G-75		0.37	12
Hypothetical Industrial HPLC Separation	Bio-Sil 250 G3000 SW	5.5 x 60	0.6	13
	Lichrosorb Diol	2.5 x 25	6.5	14

CONVENTIONAL SCALE-UP

1. Determine acceptable conditions at lab scale
2. Increase bed area proportional to the increase in feed volume, keeping all other variables constant

METHOD OF CHARM ET AL.[17]

1. Determine acceptable conditions at lab scale
2. Maintain geometric aspect ratio (L/D), Reynold's number, and feed fraction constant

Figure 1. Scale-up Methods.

To scale-up, the cross-sectional area of the bed is increased in proportion to the increase in feed volume while maintaining all other variables the same. To our knowledge, however, no systematic methodology for defining an acceptable separation has ever been put forth.

The second method, attributed to Charm et al. (17), again calls for the establishment of acceptable conditions at the laboratory scale. To scale-up, geometric aspect ratio (column length/column diameter), column Reynolds Number, and the feed loading are held constant. When we applied this method to a one-hundred fold scale-up of a typical gel chromatography system, we found that exceptionally long columns, on the order of 4-5 meters, and very slow flow rates were predicted.

Model Development

Besides these limitations, we submit that neither method is suited to optimization, they cannot be applied with confidence to all gel chromatography systems, and they reveal no knowledge of relationships among variables important to the bioprocess engineer. Consequently, we sought to develop a systematic approach to production-type gel chromatography unit operation design based on established chromatographic theory and empirical knowledge obtained experimentally.

Model Variables. Some of the variables used in this system are defined with reference to Figure 2. Shown there is a typical elution profile from a gel chromatography operation in which the earlier eluting key contaminant (subscript 1) is separated from the later eluting product (subscript 2). Protein concentration is plotted against the elution volume as a fraction of the column volume. The arrow denotes the initiation of an injection cycle, which is divided into three regimes, the feed regime, f, the run regime, r, and the wash regime, w. The elution profiles are typically described by Gaussian distribution profiles having two parameters. One parameter is the elution volume, V_e, which is analogous to the mode of a Gaussian distribution. The other, σ, is related to the width of the elution peak and is analogous to the standard deviation of a Gaussian distribution. The product is collected in the volume between the "cut" volumes, $(V_c)_1$ and $(V_c)_2$.

A production scheme for a hypothetical production-type gel chromatography operation is shown in Figure 3. Each batch having a volume of V_b, must be processed within the batch time t_b. The batch time is divided into several injections (in this case 4) and dead time during which no production occurs. As mentioned above, each injection may have three distinct flow regimes, shown here as variously shaded bars. The column must be taken out of production after n batches for cleaning and after N cleaning cycles for repacking.

In the next five figures are shown blocks from which a system for gel chromatography unit operation can be built. Each block consists of equations from chromatographic theory or empirical correlations. Each line represents either a process variable, shown entering from the left, a design variable, or a state variable, which carries information within the system. In these general

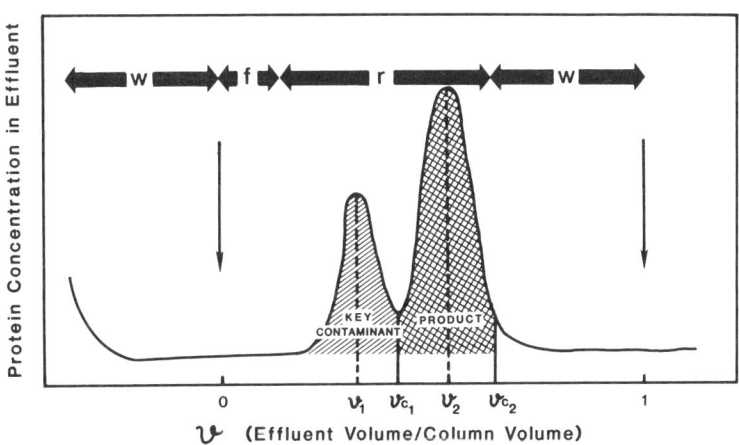

Figure 2. Separation of Product from Key Contaminant.

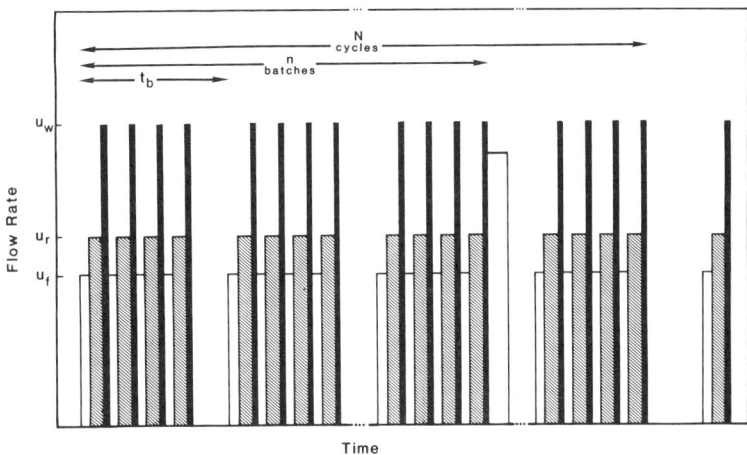

Figure 3. Sequence of Operations for a Production-Type Gel Chromatography System.

representations of the blocks, arrow heads have been omitted from the design and state variable lines because any set of the them can be solved for once the available degrees of freedom of the block are consumed.

Yield and Purity. Identification of the variables important in gel chromatography unit design begins by defining the concepts of yield and purity. These are more useful in describing the performance of production-type systems than resolution, which is useful in analytical chromatography work. The yield of product is defined as the integral of the elution concentration distribution equation of the product, assumed to be a Gaussian profile, between the cut volumes. The yield for the key contaminant is defined similarly. The purity of the product with respect to the key contaminant is defined as the ratio between the product yield and the sum of the yields of the two key solutes. The first block (Figure 4) relates the process variables, $(C_f)_i$, the concentrations of the two key solutes in the feed, the state variables, $(V_e)_i$, and σ_i, the production Gaussian distribution parameters, and the design variables, $Yield_2$, $Purity_2$, and the cut volumes, $(V_c)_1$ and $(V_c)_2$.

Loading. The production Gaussian distribution parameters can be related to those found by analytical loading of the bed by equations shown in Figure 5. In analytical loading, the feed fraction, f, is vanishingly small. The effect of large feed loads on the elution volume is given by the first equation (18). For a square wave feed (18), the second equation is derived assuming the principle of variance additivity (19).

Gaussian Parameter Correlations. In the next block (Figure 6) are equations relating the analytical Gaussian parameters to a number of design and state variables. The first equation is a rearrangement of the defining equation for the accessibility coefficient, $(K_{av})_i$ (20). The second equation is the definition of the volume of a cylinder in terms of its length and diameter. The third equation states that the total analytical variance is the sum of that due to the bed, end effect, and tubing. The bed variance in tern has been related to design variables such as flow rate, particle diameter, elution volume, and diffusivity, by correlations, such as the van Deemter (21), Giddings (15,16), or Knox (22) plate height equations. In practice, the actual form of the equation relating bed variance would have to be determined experimentally. The dependence of the variance due to the end effect on flow rate, tubing and column diameter can be determined experimentally or estimated (23). Finally, the variance caused by dispersion during flow through connecting tubing and valves can be estimated by the method of Golay and Atwood (24) and depends primarily on the tubing diameter and flow rate.

Selectivity Equation. The selectivity equation (Figure 7) relates the state variables, $(K_{av})_i$ with the process variables, mw_i, and is commonly found to be applicable over a defined molecular weight range which depends on the pore size distribution of the gel (25). The constants in the equation are dependent on the design variables, G, the gel type and P, the packing method.

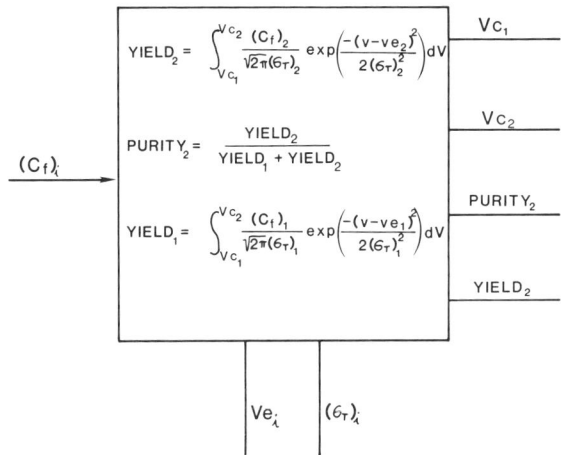

Figure 4. Yield and Purity Equations for Gaussian Peaks.

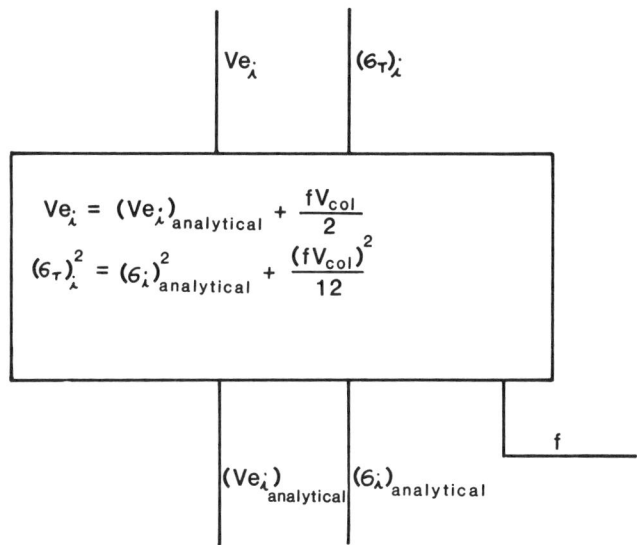

Figure 5. Feed Loading Equations.

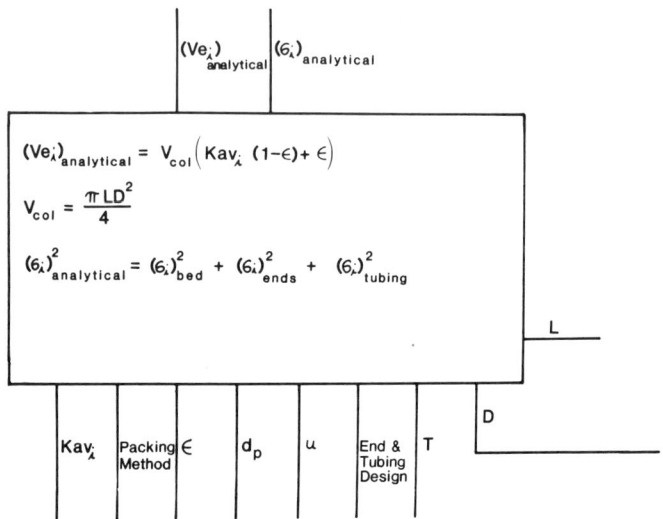

Figure 6. Gaussian Parameter Correlations.

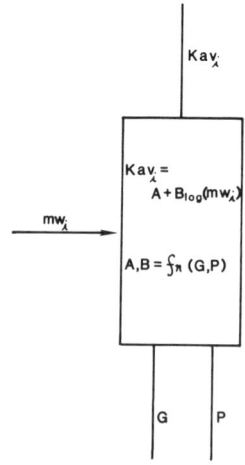

Figure 7. Selectivity Equations.

Productivity. The productivity of the process (Figure 8), expressed as grams of protein produced per unit time, is equal to the yield multiplied by the total amount of product mass entering the process, $(C_f)_2 \times V_b$, divided by the batch time t_b. This form of the equation is used when production proceeds by the sequence outlined earlier in Figure 2 and when the interuptions for bed cleaning and repacking are brief or non-existant due to the use of replacement beds.

The bed diameter equation relates the process variable, batch volume, V_b, with the design variables, f, the feed fraction, L, the bed length, i, the number of injections per batch, and D, the bed diameter as shown in Figure 9.

These blocks can be linked together to form an information flow system for a production-type gel chromatography system (Figure 10). Two additional blocks inferring relationships which have not been discussed appear in the model. One relates the packing method and gel used to the state variables, void fraction, ϵ, and effective particle diameter, d_p. The second block implies that the choice of gel constrains the flow rate and the magnitude of the efficiency loss due to extra-column causes which can be tolerated.

This system in turn can form the basis of a computer program which can be used for initial system design, such as for column sizing, or for optimization of design variables, such as feed fraction and flow rate. Following is a demonstration of how the system may be used for column sizing.

Strategy for Utilizing the Model

This complex system contains 42 variables and 25 equations. By eliminating the state variables, it can be reduced to a system of 2 independent equations with 19 process and design variables, leaving 17 degrees of freedom. In order to estimate the length and diameter of a column required for a particular separation, the degrees of freedom must be used. To do this, values for the six process variables must be assigned first. They are the concentration, $(C_f)_i$, and molecular weights, mw_i, of the two key species in the feed, the batch volume, V_b, and the batch time, t_b.

Second, values for six design variables, G,P,T,i,f, and u, plus yield and purity requirements must be specified. Good choices for some of the design variables must be made using heuristics. For example, the gel should be durable, be able to be cleaned in situ, be non-compressible, and have a high selectivity. The packing method should give reproducible, efficient, and stable packing structure for the gel used. The temperature should be the highest compatible with product and solid phase stability. The number of injections per batch should minimize the dead time with due consideration for product stability.

Two more degrees of freedom are consumed by making assumptions concerning the contributions of the column ends (23) and tubing (24) to the system variance.

The final degree of freedom is used in choosing an optimization stategy. Possible strategies include minimizing bed volume or cost/productivity.

Once the available degrees of freedom of the system have been

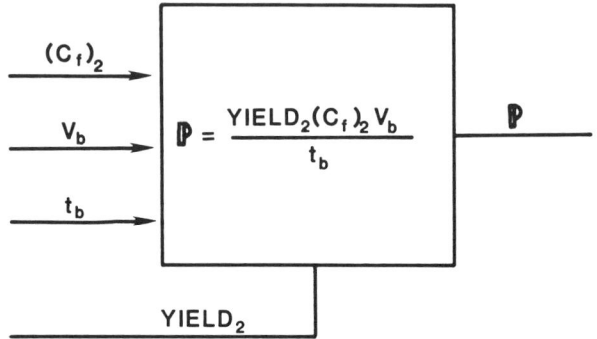

Figure 8. Productivity Equation.

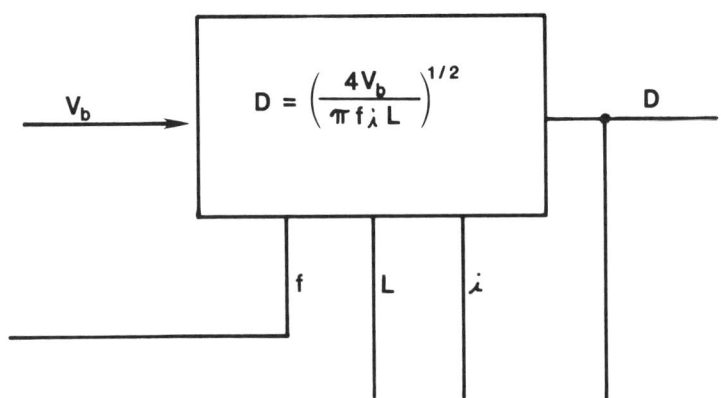

Figure 9. Bed Diameter Equation.

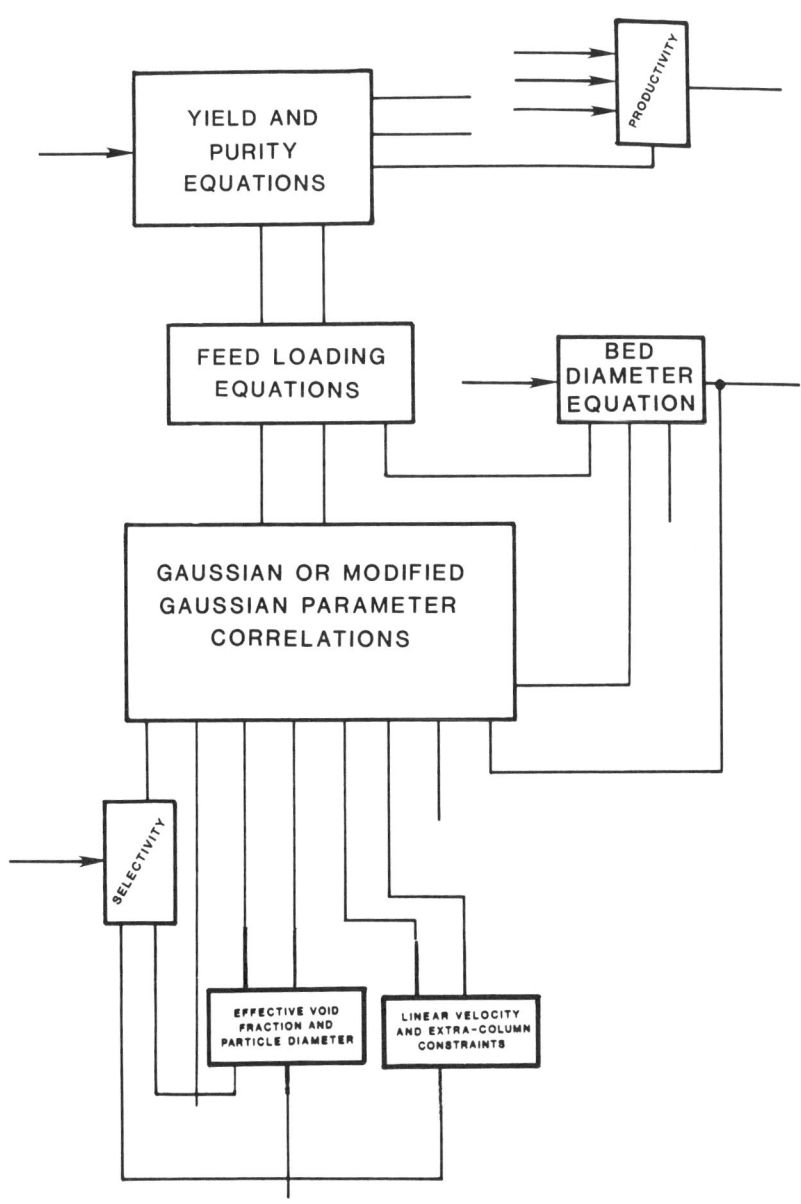

Figure 10. Information Flow Diagram for a Production-Type Gel Chromatography System.

consumed, estimates for $(K_{av})_i$ and the dependence of the bed variance on flow rate may be obtained from theoretical expressions (15,16,21,22) or from experiments at the laboratory-scale. Then, the computer program can be used to search for optimal values of column length (L) and diameter (D) as functions of each of the other design variables.

Once a column of the prescribed dimensions is obtained and set up, the actual contribution of its end effects and tubing to variance must be determined as a function of flow rate. Once the production bed is packed, the dependence of the analytical gaussian parameters on flow rate must be determined. Finally the computer program could once again be used to find feed loading and flow rates which optimize cost/productivity.

The steps in our proposed systematic approach are summarized in Figure 11. The benefits of such an approach are that production-scale systems can be more easily optimized and that scale-up can be achieved with any system. Furthermore, information which may be required by the bioprocess engineer, such as that which will signal bed deterioration or transient behavior, are obtained.

Cost Analysis

As mentioned above, there is a lack of information in the literature concerning the costs of production-type gel chromatography. Consequently, we developed an economic model for this unit operation which identified solid phase lifetime and cost and bed size as the most important variables in the model. Cost can be obtained from the manufacturer and bed size can be estimated from the program given above for any particular gel. However, literature estimates for solid phase lifetimes vary from 70 to 2000 injections (10). In order to obtain an upper estimate we made the following assumptions:
1.) Bed life of 70 injections.
2.) Gel was Sephadex G-50, superfine, operated at 10 cm/hr.
3.) Difference in molecular weights of the key species was 2-fold.
4.) Batch time was 10 hr.
5.) Feed protein concentration was 2%.
6.) Feed loading was 2%.
7.) Capital was depreciated on a straight-line basis over 10 years and included six sections of Pharmacia stack column (KS370), pumps, process control equipment, valves and tubing.
8.) Labor, filter, and chemical costs were estimated by assuming that the ratios between these costs and the cost of the solid phase were the same as those given by Curling and Cooney (26).

As a result of this exercise we estimate the total cost of an industrial-type gel chromatography operation to be on the order of $5/gram or less. This figure can be used to quickly estimate the utility of gel chromatography for a particular process. The major costs are labor (40%), gel (25%), filters (20%), chemicals (10%), and capital (5%). This estimated cost of $5/gram is quite high and, although representing an upper estimate, still indicates that gel chromatography is an expensive purification step. This estimate

1. At lab scale, obtain dependence of bed variance on flow rate and Ve of product and key contaminant

2. Use computer to estimate production-scale operating conditions

3. Once production-scale system is in place, estimate variance due to ends, tubing, valves

4. Determine dependence of bed variance on flow rate at production scale

5. Use computer to search for optimum feed load and flow rate

Figure 11. Systematic Scale-up.

is in agreement with the observation that protein fractionation by gel chromatography is only applied industrially to proteins having a high market value, such as human pharmaceuticals. It is not applied toward fractionation of bulk proteins such as whey or oilseed proteins.

Conclusion

A systematic approach for design and optimization of production gel chromatography operations which is built up from theoretical and empirical equations in the literature, and which establishes useful relationships between process and design variables has been presented. Some of the problems preventing a more widespread use of gel chromatography at the industrial scale, such as restricted productivities and poor efficiencies may be improved by the use of optimization approaches such as the one proposed. The principles are applicable to other forms of production-scale linear-isotherm chromatography carried out under constant conditions. Finally, gel chromatography represents an estimated process cost of \$5/gram product protein.

Legend of Symbols

$YIELD_i$	Proportion of component i recovered between the cut volumes relative to the amount of i applied to the column.
$PURITY_i$	Proportion of component i recovered between the cut volumes relative to the sum of the key components.
$(C_f)_i$	Concentration of component i in the feed.
D	Column Diameter
$(K_{av})_i$	Availability Coefficient = $\dfrac{(V_e)_i - V_0}{V_{col} - V_0}$
L	Column Length
N	Number of cleaning cycles between repackings.
P	Packing Method
$\mathbf{P}$	Productivity (g/h)
T	Temperature
V_b	Batch Volume
$(V_c)_j$	Cut Volume j
V_{col}	Column Volume
$(V_e)_i$	Volume of liquid effluent which leaves the column between initiation of feed and elution of the maximum concentration of i.
$(V_e)_{analytical}$	Same as $(V_e)_i$, but with vanishingly small feed volume.
V_0	Void Volume of the packed bed.
d_p	Some measurement of the diameter of packing particles.
f	Proportion of feed volume relative to column volume.
i	Number of injections of feed per batch.
mw_i	Molecular weight of i.
n	Number of feed cycles between cleanings.
r	Proportion of volume eluted during run regime to column volume.
t_b	Time required for processing of one batch.

u	Average linear velocity of liquid phase in column.
w	Proportion of volume eluted during wash regime to column volume.
ϵ	Void fraction of the packing.
$(\sigma_T)^2_i$	Variance of the elution concentration profile of i.
i	1,2 - refer to either product or key comtaninant.
j	1,2 - refer to first or second cut volume.

Literature Cited

1. Porath, J.; Flodin, P. Nature 1959, 183, 1657.
2. Dunnill, P.; Lilly, M. D. Biotechnol. & Bioeng. Symp. 1972, 3, 97.
3. Durekovic, A.; Flossdorf, J.; Kula, M. R. Eur. J. Biochem. 1973, 36, 528.
4. Bruton, C.; Jakes, R.; Atkinson, T. Eur. J. Biochem. 1975, 59, 327.
5. Bascomb, S.; Banks, G. T.; Skarstedt, M. T.; Fleming, A.; Bettelheim, K. A.; Connors, T.A. J. Gen. Microbiol. 1975, 91,1.
6. Janson, J. C. J. Agr. Food Chem. 1971, 19, 581.
7. Curling, J. M. Amer. Lab. 1976, 8, 26.
8. Hanford, R.; Maycock, W.; Vallet, L. in "Chromatography of Synthetic and Biological Polymers", vol. 2, R. Epton, Ed., Ellis Harwood, Chichester, U.K. 1978, pp. 111-119.
9. Horton, T. Amer. Lab. 1972, 4, 83-91.
10. "The large-Scale purification of Insulin by gel filtration chromatography", Pharmacia, Uppsala, 1983.
11. Donnelly, E. B.; Delaney, R. A. M. J. Food Technol. 1977, 12, 493.
12. Nichols, R. A.; Fox, J. B. J. Chromatogr. 1969, 43, 61.
13. BIO-RAD Technical Information. 1985. BIO-RAD Laboratories. pp.111-114.
14. Roumeliotis, P.; Unger, K. J. Chromatogr. 1979, 185, 445.
15. Giddings, J. C. "Dynamics of Chromatography, Part 1, Principles and Theory", Dekker, New York, 1965.
16. Giddings, J. C.; Mallik, K. L. Anal. Chem. 1966, 38, 997.
17. Charm, S. E.; Matteo, C. C.; Carlson, R. Anal. Biochem. 1969, 30, 1.
18. Personnaz, L.; Gareil, P. Sep. Sci. Technol. 1981, 16, 135.
19. Sternberg, J. C. Adv. Chromatogr. 1966, 2, 205.
20. Kremmer, T.; Boross, L. "Gel Chromatography, Theory, Methodology, Applications", John Wiley & Sons. Chichester. 1979.
21. van Deemter, J. J.; Zuiderweg, F. J.; Klinkenberg, A. Chem. Eng. Sci. 1956, 5, 271.
22. Knox, J. H.; Saleem, M. J. Chromatogr. Sci. 1969, 7, 614.
23. Coq, B.; Cretier, G.; Rocca, J. L.; Porthault, M. J. Chromatogr. Sci. 1981, 19, 1.
24. Golay, M. J. E.; Atwood, J. G. J. Chromatogr. 1979, 186, 353.
25. Ouano, A. C.; Barker, J. A. Sep. Sci. 1973, 8, 673.
26. Curling, J. M.; Cooney, J. M. J. Parent. Sci. Technol. 1982, 36, 59.

RECEIVED April 16, 1986

15

Electron Paramagnetic Resonance Spectroscopy Studies of Immobilized Monoclonal Antibody Structure and Function

Erik J. Fernandez, Forrest B. Fernandez, Roger B. Jagoda, and Douglas S. Clark[1]

School of Chemical Engineering, Cornell University, Ithaca, NY 14853

> Electron paramagnetic resonance (EPR) spectra were taken of two spin labeled haptens combined with soluble and immobilized anti-2,4-dinitrophenyl (anti-DNP) IgG_{2b} and IgE antibodies. Different association constants between soluble antibodies and spin labeled haptens were accompanied by differences in maximum peak-to-peak splittings in EPR spectra. When the IgG_{2b} antibodies were immobilized using CNBr-activated Sepharose and immobilized Protein-A, lowered specific activities were observed (1.2 for CNBr-Sepharose at .58 mg antibody/ml gel, 0.8 for Protein-A Sepharose at 1.7 mg antibody/ml gel), and the binding constant of CNBr Sepharose-immobilized antibody was 44% that of the soluble antibody. Furthermore, the EPR spectrum of spin label combined with IgG_{2b} changed measurably upon immobilization of the antibody to CNBr-Sepharose.

Affinity chromatography utilizing monoclonal antibodies immobilized to a solid matrix has been used as a purification method for several years with characteristic advantages and disadvantages.(1-3) Antibodies are proteins produced by vertebrates "designed" to bind specific molecules, referred to as antigens. When utilized in the laboratory or production facility, antibodies immobilized to a suitable support allow an antigen to be separated from many other components of a biochemical process stream. The strong interaction between antibody and antigen (with equilibrium constants as high as $10^{14} M^{-1}$) (4) can be a great advantage in product recovery from fermentation broths, where product concentrations can be as low as 10^{-6} g/ℓ in extreme cases.(5) It can also be a disadvantage, however, since potentially denaturing conditions may be required to

---

[1]To whom correspondence should be addressed

cause dissociation of a tightly bound antigen. Another disadvantage of immunoaffinity chromatography is the cost. Monoclonal antibodies purchased in large quantities from specialized companies now sell for about $2000/g, although this number is expected to drop significantly as more cost effective production methods are developed. Nonetheless, a wide variety of biological products have been purified using monoclonal antibodies, at least in the laboratory. (For an overview, see reference 6).

In order to maximize the utility of immobilized monoclonal antibodies and minimize costs when they are used, it is necessary to have a molecular-level understanding of the effects of immobilization on antibodies used in the preparation of immunosorbents. Several studies have already shown that immobilization can have a significant effect on antibody activity. For example, Bolton and Hunter (7) found that the total activity of anti-human growth hormone antibodies fell when coupled to Sepharose and cellulose supports. In addition, they observed a drop in what they called "sensitivity", a quantity inversely related to the binding strength of the antibody. In a related study, Eveleigh and Levy (8) studied the effect of immobilized antibody loading on specific activity, that is, the number of antigens that could bind to an antibody molecule. They found that as the loading of antibody on the support was increased, the overall binding capacity of the immunosorbent increased, but the specific activity of immobilized antibody dropped. As a final example consider the results of Weston and Scorer, who found that as the loading of IgG antibody on Sepharose was increased, the specific activity of the antibodies decreased to the extent that a maximum in total binding capacity was observed.(9) Before postulating probable causes of this behavior, it is helpful to review some of the general characteristics of antibodies.

Antibodies, or immunoglobulins, are globular proteins with molecular weights of 150 to 200 kdal. They are made up of four sub-units, two "heavy" and two "light" chains, which together form a "Y"-shaped molecule. The stem of the antibody is referred to as the "constant" or F_c region, while the two branches containing the combining sites are called F_{ab} fragments. Notably, these two regions are joined by few covalent bonds, giving the molecule a great deal of motional flexibility.

With these structural features in mind, there are several possible causes one might put forward as responsible for the reduced specific activity of antibodies on a support, as depicted schematically in Figure 1. First, a portion of antibodies covalently bound in random orientations can be attached via linkages close to a combining site, effectively blocking it. Second, antibody-antibody steric effects could hinder the ability of an antigen to bind to a combining site, thereby reducing the number of "active" sites. Third, lowered specific activity is expected for antibody molecules immobilized in regions of the support either partially or completely inaccessible to antigens. And finally, a different type of inactivation will result if adverse conformational changes occur as a result of the immobilization process.

The aim of this research, then, is to better understand which of these mechanisms is or are important in formulating immunosorbents of optimal activity and binding strength. To do this it is necessary to determine the activity and conformation of antibodies immobilized to a support. Unfortunately, the supports often used (e.g. agaroses) interfere with most direct physical measurements of protein structure and function. One method largely insensitive to the nature of the support, however, is EPR spectroscopy.

Materials and Methods - EPR Spectroscopy

EPR, or electron paramagnetic resonance spectroscopy, is a magnetic resonance technique designed to detect species containing unpaired electrons.(10-11) It is similar in principle to NMR, the primary difference being that EPR is based on the magnetic moments of unpaired electrons rather than the magnetic moments of specific nuclei. Since antibodies don't normally possess any unpaired electrons, small antigens, i.e. haptens, containing stable free radicals ("spin labels") can be used along with EPR to probe the antibody combining site. By utilizing the two different 2,4-dinitrophenyl (DNP) spin labels shown in Figure 2, we are investigating the conformation and activity of anti-DNP monoclonal antibodies in solution and immobilized by different methods. EPR spectroscopy is a powerful method for this kind of study because the EPR spectrum can provide detail about the motion of the spin label and thus the conformation of the antibody combining site. In this way, EPR can help to elucidate which of the possible inactivation mechanisms discussed above is or are important.

Shown in Figure 3 are EPR spectra of spin label FDNP-SL in solution and bound to anti-DNP antibody. The more restricted motion of antibody-bound antigen results in broader lineshapes and a larger maximum peak-to-peak splitting, termed A_{max}. The three-peaked spectrum of spin label in solution, for instance, is characterized by an A_{max} of about 34G, whereas the spectrum of antibody-bound FDNP-SL has an A_{max} of 61G. It should also be noted that the antibody bound spectrum is a composite spectrum containing both the characteristic three-peaked free spectrum and a bound spin label spectrum of broader lineshape. Since the signal intensity of each spectrum is proportional to the concentration of spin label, the heights of isolated peaks and/or the double integral of an entire absorption spectrum can be used to determine concentrations of free and bound spin label.(12) For instance, in the case of the free label, either the peak height or double integral can be used. In the case of the composite spectrum, however, the integral provides the concentration of both free and bound label, and it is necessary to subtract the concentration of free label from the total to determine the amount of spin label bound to antibody. By doing so for samples containing different ratios of free to bound spin label, the equilibrium constant between antibody and hapten can be obtained directly from EPR spectra.

A more sophisticated analysis of spin label motion in the combining site involves the analysis of experimental EPR spectra with the aid of computer simulations, such as those of Freed *et al.*

15. FERNANDEZ ET AL. *EPR Spectroscopy of Monoclonal Antibodies* 211

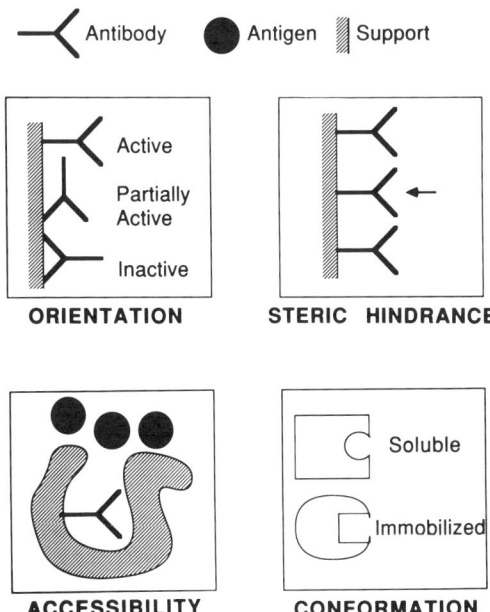

Figure 1. Causes of inactivation of immobilized antibodies.

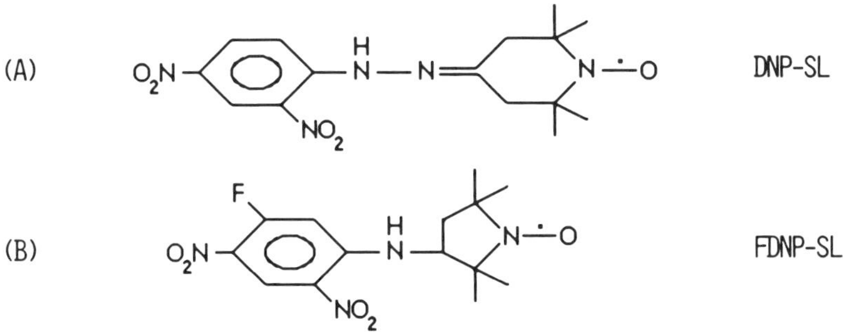

(A) O$_2$N—⟨⟩—N(H)—N=⟨piperidine-N—O⟩   DNP-SL
          NO$_2$

(B) O$_2$N—⟨F⟩—N(H)—⟨pyrrolidine-N—O⟩   FDNP-SL
          NO$_2$

Figure 2. Spin labels used: (A) DNP-SL, (B) FDNP-SL.

(13) By comparing simulated spectra with those obtained experimentally, it is possible to distinguish between different models of the motion of a spin label. (14) The primary motional parameters involved are "rotational correlation times", characteristic times of rotational diffusion around each of the three axes of a coordinate system defined with respect to the spin label. As shown in Figure 4a, the x-axis is defined as colinear with the N-O bond of the nitroxide portion of the molecule, the z-axis as perpendicular to the "plane" of the ring, and the y-axis as perpendicular to the x-z plane. Efforts to match computer-simulated and experimental EPR spectra revealed that the best model incorporates a "diffusional tilt" of angle β with respect to the molecular coordinate system, along with three rotational correlation times, τ_x, τ_y, and τ_z, as shown in Figure 4b.

Results and Discussion

The best fit to date between simulated and experimental spectra was obtained with soluble IgG_{2b}, with β equal to 50°, τ_x = 5.5 ns, and τ_y and τ_z = 33 ns. This represents a fast rotation of the spin label around the x' axis and slower motion around the other axes as depicted in Figure 4b. It is interesting to note that 33 ns agrees well with the rotational correlation time of 47 ns for the F_{ab} fragment of an IgE molecule determined by Slattery, et al.(15) using steady state anisotropy measurements. This agreement suggests that τ_y and τ_z represent time constants for rotational motion of the F_{ab} fragment as a whole. Unfortunately, because of the complex and time-consuming nature of the simulations, nothing further can be reported at this time about the motion of the spin label in the combining site through correlation times and tilt angles. However, there is also useful information contained in maximum peak-to-peak splittings.

In studies to date, different spin labels have produced different spectra when combined with different antibodies. For example, the splittings and binding constants for the two different spin labels binding to soluble anti-DNP IgG_{2b} antibodies are summarized in Table I. These data show that EPR is sensitive to differences in the motion of (1) the same spin label in different motional environments (i.e., different antibody combining sites),

Table I. Maximum peak-to-peak splittings and equilibrium binding constants for soluble IgG_{2b} and IgE combined with DNP-SL and FDNP-SL.

Antibody	Spin Labeled Hapten	K (M^{-1})	A_{max} (gauss)
IgG_{2b}	DNP-SL	2.2×10^6	62
IgG_{2b}	FDNP-SL	3.9×10^6	56
IgE	DNP-SL	1.3×10^7	61
IgE	FDNP-SL	4.8×10^6	≤ 56

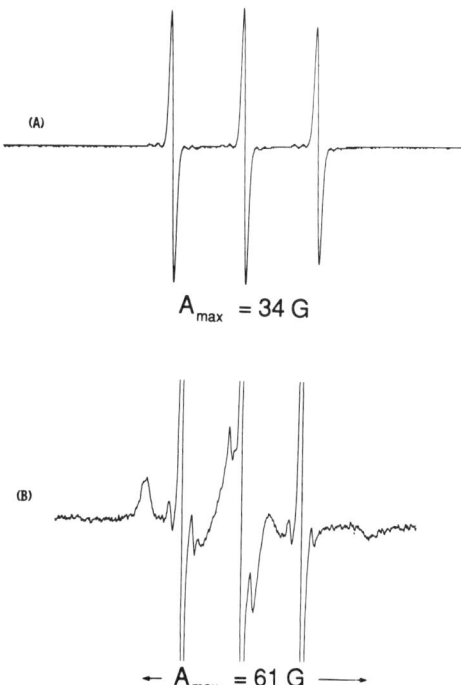

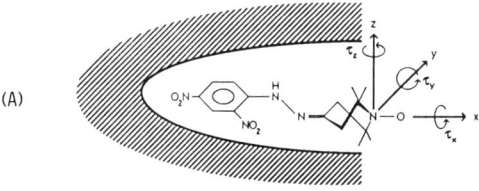

Figure 3. EPR spectra of (A) free and (B) bound spin label.

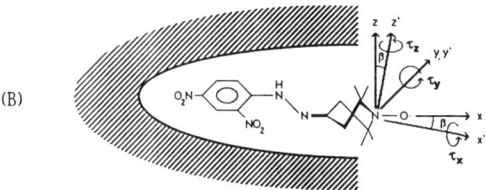

Figure 4. Schematic of spin label in antibody combining site: (A) without "diffusional tilt", (B) with "diffusional tilt".

and of (2) different spin labels in the same environment (i.e., the same antibody combining site). This suggests that EPR might be able to discriminate motional differences caused by binding site conformational changes. If this is the case, EPR should be a useful tool in studying structure-function relationships in immobilized antibodies as well.

For immobilization studies to date, two distinct modes of immobilization have been used. The first utilizes nonspecific covalent bonding to CNBr-activated Sepharose via primary amino groups on the antibody molecule. Since there are many of these available on the antibody, this is expected to result in random orientation of antibody molecules on the support. The other method involves linkage through immobilized protein A, a protein which binds immunoglobulins in the structural F_c portion of the molecule. (15) Because this method should result in immobilized antibody molecules with more optimal orientations, it is expected to produce samples with higher activity, although it would certainly be a more expensive and complex technique to carry out on a large scale.

Shown in Table II are activity and binding constant data for samples of immobilized antibodies prepared by the two different methods. The loading of IgG on the support was determined by amino acid analysis, and the amount of active and accessible antibody was determined by EPR spectroscopy. The binding constant for the CNBr-Sepharose sample was determined by fluorescence titration.(16) Both samples have lowered specific activity with respect to the ideal value of 2.0 (remember that there are two combining sites per antibody molecule), and the binding constant for the CNBr-Sepharose sample is only 44% that of its value in solution. It is also very interesting to note that contrary to expectations the Protein-A sample has a lower specific activity, which may be due to the fact that it has a much higher loading than the CNBr-Sepharose sample. Current research in this laboratory should soon provide a more definitive explanation for these results.

Table II. Loadings, specific activities, and binding constants for samples of immobilized IgG_{2b}.

Support	Coupled IgG by Amino Acid Analysis (mg/ml)	Active/Accessible IgG by EPR (mg/ml)	Specific Activity	K (1/M)
CNBr-Sepharose	0.58	0.35	1.2	1.7×10^6 (44%)
Protein A-Sepharose	1.7	0.70	0.8	?

Shown in Figure 5 are EPR spectra of FDNP-SL combined with IgG_{2b} in solution and immobilized to CNBr-Sepharose. Although an accurate A_{max} cannot be determined from these particular spectra, the left-most peak in the spectrum of immobilized antibody is shifted significantly with respect to its position in the spectrum of antibody in solution. This shift indicates that the overall mobility of the spin label is more restricted when the label occupies the combining site of immobilized antibody. The accompanying decrease in the binding constant observed upon immobilization indicates that such changes in spin label mobility are due at least in part to changes in combining site conformation; however, altered motion of the entire F_{ab} fragment may also have influenced the spectrum. Clarification of this point, provided by the use of computer simulations to determine rotational correlation times for the immobilized antibody system is expected in the near future.

Concluding Remarks

The data presented above establish the applicability of EPR to structure-function characterization of immobilized antibodies. The soluble A_{max} determinations indicate that EPR is sensitive to different environments and to the structure of the probe used. They also suggest that different probes might be used to obtain a more complete picture of the conformation of the binding site. In addition, the measurable differences between the EPR spectra of soluble and immobilized antibodies indicate that immobilization has significantly affected the motion of the spin label in the combining site. Although the data presented here do not allow specific conclusions to be drawn about the effect of immobilization on antibody conformation and activity, they do illustrate that EPR can provide molecular level insights not available by traditional methods. Thus, EPR spectroscopy should serve as a powerful tool

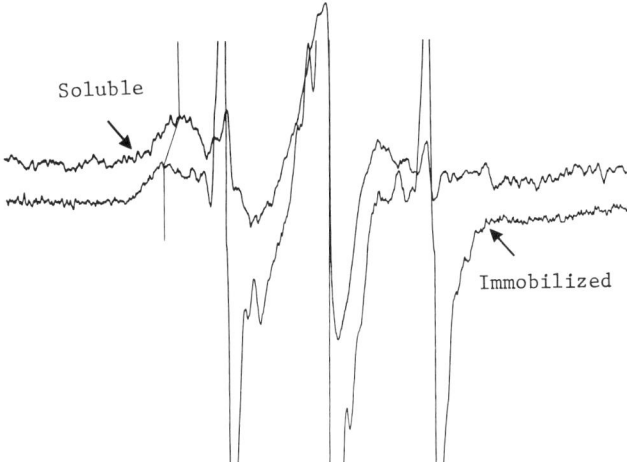

Figure 5. EPR spectra of soluble and CNBr Sepharose-immobilized IgG_{2b}.

for direct experimental analyses of immobilized antibody structure and function, and therefore help to elucidate the basis for the suboptimal binding capacities frequently exhibited by immunosorbents.

Acknowledgments

The authors are indebted to Dr. Barbara Baird, D. David Holowka, and Dr. Normal Klinman for their assistance in obtaining antibodies and to Dave Schneider and Glenn Millhauser for their advice on the spectral simulations. This material is based upon work supported under a National Science Foundation Graduate Fellowship.

Literature Cited

1. Turkova, J. In "Affinity Chromatography"; Elsevier: New York, 1975; p. 95.
2. Chase, H. A. Chem. Eng. Sci. 1984, 39, 1099.
3. Goding, J. W. In "Monoclonal Antibodies: Principles and Practice"; Academic: New York, 1983; pp. 188-207.
4. Mukkur, T. K. S. Biochem. J. 1978, 172, 39.
5. Dwyer, J. W. Bio/Technology 1984, 2, 957.
6. Low, D. "Proceedings of Biotech 83, International Conference on the Commercial Applications and Implications of Biotechnology"; Online: Northwood, U.K. 1983, p. 830.
7. Bolton, A. E.; Hunter, W. M. Biochim. Biophys. Acta 1973, 329, 318.
8. Eveleigh, J. W.; Levy, D. E. J. Solid-Phase Biochem. 1977, 2, 45.
9. Weston, P. D.; Scorer, R. In "Affinity Chromatography"; Hoffman-Ostenhof, et al., Ed.; Pergamon: New York, 1978, p. 207.
10. Schumacher, R. T. "Magnetic Resonance"; W. A. Benjamin: New York, 1970.
11. Cantor, C. R.; Schimmel, P. R. In "Biophysical Chemistry"; W. H. Freeman: New York, 1980; pp. 525-536.
12. Bailey, J. E.; Clark, D. S. In "Methods in Enzymology"; Mosbach, K., Ed.; Academic: New York, in press.
13. Meirovitch, E.; Igner, D.; Igner, E.; Moro, G.; Freed, J. H. J. Chem. Phys. 1982, 77, 3915.
14. Goldman, S. A.; Bruno, G. V.; Polnaszek, C. F.; Freed, J. H.; J. Chem. Phys. 1972, 56, 716; Hwang, J. S.; Mason, R. P.; Hwang, L. P.; Freed, J. H. J. Phys. Chem. 1975, 79, 289; J. Phys. Chem. 1975, 79, 2283; Campbell, R. F.; Freed, J. H. J. Phys. Chem. 1980, 84, 2668; Meirovitch, E.; Freed, J. H. J. Phys. Chem. 1980, 84, 2459; Shiotani, M.; Moro, G.; Freed, J. H. J. Chem. Phys. 1981, 74, 15.
15. Slattery, J.; Holowka, D. A.; Baird, B. A. Biochemistry, in press.
16. Goding, J. W. In "Monoclonal Antibodies: Principles and Practice"; Academic: New York, 1983; p. 195.
17. Mukkur, T. K. S.; Szewczuk, M. R.; Schmidt, E. E., Jr. Immunochem. 1974, 11, 9.

RECEIVED April 16, 1986

INDEXES

Author Index

Asenjo, J. A., 9
Bier, Milan, 185
Chang, Ho Nam, 32
Chung, Bond Hyun, 32
Clark, Douglas S., 208
Dove, G. B., 93
Drioli, Enrico, 52
Ercoli, E., 43
Fernandez, Erik J., 208
Fernandez, Forrest B., 208
Finn, R. K., 43
Fisher, R. R., 109
Glatz, C. E., 109
Gobie, William A., 169
Hatton, T. A., 67
Hettwer, David J., 2

Hunter, J. B., 9
Ivory, Cornelius F., 169
Jagoda, Roger B., 208
Kaul, Rajni, 78
Kelley, James J., 193
Kim, In Ho, 32
Ladisch, M. R., 122
Mattiasson, Bo, 78
Mitra, G., 93
Nigam, Somesh C., 153
Rudge, S. R., 122
Thien, M. P., 67
Wang, D. I. C., 67
Wang, George Y., 193
Wang, Henry Y., 2,153,193

Subject Index

A

Acetic acid production, extractive fermentation in aqueous two-phase systems, 80
Acetone-butanol production, extractive fermentation in aqueous two-phase systems, 80
Adsorbent, immobilized affinity, mathematical modeling of bioproduct adsorption, 153
Adsorbent bead, immobilized affinity, schematic diagram, 156f
Adsorbent matrix, effect of bioproduct diffusivity on ligand consumption, 164f
Aerobic whole-cell immobilization, dual hollow-fiber bioreactor, 32
Affinity adsorbent, immobilized bead, schematic diagram, 156f
 hydrogel beads, 158-164
 mathematical modeling of bioproduct adsorption, 153
Affinity adsorption, theory, 154-158
Affinity chromatography, utilizing immobilized monoclonal antibodies, 208
Affinity partitioning
 aqueous two-phase systems, 86t
 proteins, aqueous two-phase systems, 85-87
Affinity sorbent, partitioning in two-phase system, 88f

Affinity support, scale-up of chromatographic processes, 124
Aggregate growth and breakup, protein precipitation, 112
Agitation speed, effect on LEM separation, 73f
Albumin
 concentration and partition coefficient as function of salt concentration, PEG-water solution, 99f
 protein recovery by salt partition, 107
Alcoholic fermentations, immobilized systems and aqueous two-phase systems, 80
Alcohol, ion exchange separation, 125f
Allergens, RIEF separation, 187,188f
Allosteric enzymes, gelled membrane formation, immobilized, 61
Amino acids
 membrane reactor, 58
 process diagram for LEM-based recovery, 76f
 recovery from lysate, membrane process, 56
 recovery using liquid emulsion membranes, 71-75
Amphoteric compounds, IEF, 186
Amylase and cellulase, aqueous two-phase systems, semicontinuous production, 84
Anaerobic single-cell protein (SCP), membrane reactor, 43,44

Analysis of cell components, protein
  release from E. coli, 3
Analysis of spin label motion,
  monoclonal antibodies, 210
Antibiotics, concentration and
  purification by sequential
  UF and RO, 56
Antibodies
  description, 209
  EPR spectroscopy studies of
    structure and function,
    immobilized monoclonal, 208
  materials and methods, EPR
    spectroscopy, 210
  simulated and experimental spectra,
    EPR, immobilized monoclonal, 212
Antigens, aqueous two-phase system
  production, 84
Aqueous multiphase systems, phase
  partitioning, 95
Aqueous two-phase systems
  affinity partitioning, 86t
  alcoholic fermentations compared to
    immobilized systems, 80
  biocatalytic reaction, 79
  bulk chemical production, 82t
  economy, 89
  equipment for continuous
    extraction, 90
  extractive bioconversion, 81f
  fine chemicals production
    compared with immobilized
    systems, 83
  discussion, 82
  isolation of enzymes, 85t
  macromolecules production, 84
  product inhibition, fermentation
    processes, 79-85
  purification of proteins by
    partitioning, 85-87
  recovery and purification in
    biotechnology, overview, 78-90
  recovery of proteins by salt
    partitioning, 93
Aspergillus niger
  cultivation in dual hollow-fiber
    bioreactor, 37
  immobilization in dual hollow-fiber
    bioreactor, 32
Asymmetric enzyme capillary membranes,
  phase inversion method, 63f

Binding
  single-component diffusion,
    immobilization in
    hydrogel, 158-163
  two-component diffusion,
    immobilization in hydrogel, 163
Bio-affinity, scale-up of
  chromatographic processes, 124
Biocatalytic reaction, aqueous
  two-phase system, 79
Biochemical separations, LEMs,
  commodity-type, 70
Bioconversion, aqueous two-phase
  systems, extractive fermentation
  processes, 79-85
Biological materials, colloidal,
  electrokinetic parameters, 180t
Bioreactor
  advantages and disadvantages,
    hollow-fiber membrane, 32
  aerobic whole-cell immobilization,
    dual hollow-fiber, 32,33
  continuous production of rifamycin B,
    dual hollow-fiber, 40f
  cross sectional view of dual
    hollow-fiber, 34f
  cultivation of A. niger,
    deformed, 39f
  cultivation of E. coli, dual
    hollow-fiber, 35
  electron micrographs of densely
    packed E. coli, dual
    hollow-fiber, 38f
  schematic diagram of experimental
    setup, dual hollow-fiber, 36f
Biosurfactants, aqueous two-phase
  systems, extractive conversion, 83
Biotechnology
  definition, downstream
    processing, 52
  overview, aqueous two-phase systems
    for recovery and
    purification, 78-90
  summary of membrane processes, 55t
Bonding, immobilization methods,
  monoclonal antibodies, nonspecific
  covalent, 214
Breakage models, protein
  precipitation, 116
Bulk chemical production, aqueous
  two-phase systems, 80,82t
Buoyancy, electrophoresis scale-up
  considerations, 138

B

C

Bed diameter equation, gel
  chromatography design model, 202f

Calculated elution profile
  chromatographic protein
  separation, 137f

Calculated elution profile—Continued
　electrochromatographic protein
　separation, 145
Carbohydrates, soluble protein and
　peptides, enzymatic lysis and
　disruption of yeast cells, model
　simulations, 21-24
Cell, microbial, subcellular location
　of enzyme activities, 10
Cell breakage, mechanical disruption
　of E. coli, 5f
Cell components, protein release from
　E. coli, 3
Cell concentration, effect on protein
　release profile, 7f
Cell fractionation, model, enzymatic
　lysis and disruption of yeast
　cells, 24
Cell fragmentation, mechanically based
　protein release methods, 2
Cell immobilization, dual hollow-fiber
　bioreactor, aerobic, 32,33
Cell lysis, sequence, enzymatic lysis
　and disruption of yeast cell, 11
Cell permeabilization, protein release
　from E. coli, 3
Cell preparation, protein release from
　E. coli, 3
Cellular protein release
　chemical treatment of E. coli, 4,5f
　mechanical disruption of E. coli, 5f
Cellulase, amylase and, semicontinuous
　production in aqueous two-phase
　systems, 84
Cellulose, enzymatic degradation to
　alcohol, membrane reactor, 58
Chemical permeabilization, protein,
　DNA, and RNA release from
　E. coli, 4
Chemicals
　bulk, production in
　　two-phase systems, 80,82t
　extractive fermentation in two-phase
　　systems, 80
　fine, production in aqueous
　　two-phase systems, 82
Chloride countertransport system,
　typical LEM separation, 73f
Chromatographic apparatus for process
　system, 127f
Chromatographic processes, scale-up
　considerations, 126-129
Chromatographic protein fractionations
　calculated elution profile, 137f
　gel, industrial and large
　　scale, 195t
Chromatographic supports, scale-up of
　chromatographic processes, ion
　exchange resins, 124
Chromatography
　affinity, immobilized monoclonal
　　antibodies, 208

Chromatography—Continued
　gel
　　model development, 196-201
　　purification of protein products,
　　　large-scale, 193
　　scale-up processes, theoretical
　　　considerations in size
　　　exclusions, 129-135
　　size exclusion, equilibrium
　　　expressions and mass transfer
　　　expressions, 131t
Colloidal and biological materials,
　electrokinetic parameters, 180t
Column area, chromatographic scale-up
　considerations, 126
Column length, chromatographic
　scale-up considerations, 128
Commodity-type biochemicals,
　separation, LEMs, 70
Concentration profile of
　cycloheximide, immobilized
　adsorbent beads, 160f
Continuous-flow electrophoresis (CFE)
　classic, schematic, 172f
　classic thin-film, general
　　discussion, 170
　comparison of single-pass
　　models, 175f
　discussion, recycle, 169,171-173
　general discussion, 170
　schematic, classic, 172f
Continuous production of rifamycin B,
　dual hollow-fiber bioreactor, 40f
Convective dispersion, RCFE,
　model, 174
Conversion of biosurfactants, aqueous
　two-phase systems, extractive, 83
Cost analysis, gel chromatography
　scale-up, 204
Countertransport system
　LEM separation, chloride, 73f
　LEM-mediated amino acid recovery, 71
Covalent bonding, nonspecific,
　monoclonal antibody immobilization
　methods, 214
Cross-flow microfiltration, bioactive
　compounds from fermentation
　broths, 52
Cycloheximide, immobilized adsorbent
　beads, concentration profile, 160f

D

Deformed bioreactor, cultivation of
　A. niger, 39f
Diffusion and binding
　single component, immobilization in
　　hydrogel, 158-163
　two component, immobilization in
　　hydrogel, 163

Dimensionless parameters, RCFE
  model, 176t
Dispersion coefficient, effective,
  electrophoretic mobility,
  RCFE, 175f
DNA release
  E. coli treated with guanidine HCl
    and Triton, 5f
  mechanically disrupted E. coli, 4
Double-layered structure of the yeast
  wall, 12f
Downstream processing
  biotechnology, definition, 52
  membrane systems, 53-57
  phenylalanine from fermentation
    broth, LEM-mediated
    separations, 71
Drug delivery, LEMs, 70
Dual hollow-fiber bioreactor
  aerobic whole-cell
    immobilization, 32,33
  electron micrographs of densely
    packed E. coli, 38f
  schematic diagram of experimental
    setup, 36f

E

Effective dispersion coefficient,
  electrophoretic mobility,
  RCFE, 175f
Effective osmolality of cell lysate,
  enzymatic lysis and disruption of
  yeast cells, 16
Electrochromatography
  process considerations for
    scale-up, 122
  protein separation, calculated
    elution profile, 145
  schematic diagram, 140f
Electrodialysis (ED)
  bioactive compounds from
    fermentation broths, 52
  summary, biotechnology, 55t
Electrokinetic parameters of colloidal
  and biological materials, 180t
Electrokinetic separations,
  electrophoresis, scale-up
  considerations, 143
Electron micrograph
  densely packed E. coli, dual
    hollow-fiber bioreactor, 38f
  densely packed N. mediterranei, dual
    hollow-fiber bioreactor, 41f
Electron paramagnetic resonance (EPR),
  immobilized monoclonal
  antibodies, 208,210,212
Electroosmosis, scale-up
  considerations,
  electrophoresis, 136

Electrophoresis
  continuous flow
    general discussion, 170
    recycling effluent, 169
    factors affecting electrophoretic
      mobilities, 135
    first free-flow device, 169
    production-scale, center for
      separation science, 191
  scale-up considerations
    controlled convection, 138
    discussion, 122,135
    electroosmosis, 136
    isoelectric focusing, 139
    joule heating, viscosity, and
      buoyancy, 138
    plate height and resolution
      concepts, 141-143
Electrophoretic mobility
  factors affecting, 135
  RCFE, effective dispersion
    coefficient, 175f
Elution profile
  chromatographic, protein
    separation, 137f
  electrochromatographic, protein
    separation, 145
  size exclusion chromatography, 132
Enzymatic degradation of cellulose to
  alcohol, membrane reactor, 58
Enzymatic hydrolysis of triglyceride,
  membrane reactor, 58
Enzymatic lysis and disruption of
  yeast cells, model
  applications, 9,24
Enzymatic-microbial methods, fine
  chemical production, 83
Enzyme
  accumulation, enzymatic lysis and
    disruption of yeast cells, 24
  isolation in aqueous two-phase
    systems, 85t
  lytic system
    enzymatic lysis and disruption of
      yeast cell, 11
    protein release from yeast
      cells, 10
  partitioning, multiphase aqueous
    systems, 96
  release of site specific by yeast
    lysis, 25f
  RIEF, 191
Enzyme membrane reactors
  asymmetric capillary, phase
    inversion method, 63f
  discussion, 58
  formation, S. solfataricus,
    gelled, 61-65
  research and development, 59
Enzyme recovery from subcellular
  structures, yeast lysis, 28f
Enzyme release simulation, yeast lysis
  model, process conditions, 26

INDEX

Equilibrium expressions, size
  exclusion chromatography, 131t
Escherichia coli
  cultivation, dual hollow-fiber
    bioreactor, 35
  electron micrographs, dual
    hollow-fiber bioreactor, 38f
  immobilization in dual hollow-fiber
    bioreactor, 32
Ethanol, extractive fermentation in
  aqueous two-phase systems, 80
Extraction, continuous, aqueous
  two-phase systems, 90
Extractive bioconversion
  biosurfactants, aqueous two-phase
    systems, 83
  principle, aqueous two-phase
    systems, 81f
  product inhibition, aqueous
    two-phase systems, 79-85
Extractive fermentation, chemicals,
  two-phase systems, 80

F

Facilitated transport
  LEMs, 68
  phenylalanine-chloride system,
    mechanism, 69f
Feed loading equations, gel
  chromatography design model, 199f
Fermentation broth, LEM-mediated
  separations, downstream processing
  of phenylalanine, 71
Fermentation chambers, simultaneous
  production of SCP and methane, 46
Fermentation processes, extractive
  bioconversions, product
  inhibition, 79-85
Fermentation products
  RIEF, 191
  total market values, 54t
Fine chemicals, production in aqueous
  two-phase systems, 82
Fractionation
  industrial and large scale, gel
    chromatographic protein, 195t
  RIEF
    illustrative, 187-189
    large-scale protein, 185

G

Gaussian parameter correlations, gel
  chromatography design
  model, 198,200f

Gel chromatography
  current design methods, 194
  design model, use, 201-204
  model development, 196-201
  protein fractionation
    industrial and large scale, 195t
    scale-up methods, 195f
  protein purification
    inherent problems, 194
    large-scale, 193
    scale-up cost analysis, 204
    sequence of operations,
      production-type, 197f
Gelled enzyme membrane formation,
  S. solfataricus, 61-65
Glucan hydrolysis rate, enzymatic
  lysis and disruption of yeast
  cells, 16
Glucose
  ion exclusion separation, 127f
  and pH histories, effluent, dual
    hollow-fiber bioreactor, 36f
Glycerine and acids, membrane
  reactor, 58
Gradients, pH, RIEF, 189-191
Guanidine HCl and Triton, synergistic
  effect on protein release profile,
  E. coli, 4,6f

H

Hollow-fiber bioreactor
  dual
    aerobic whole-cell
      immobilization, 33
    continuous production of
      rifamycin B, 40f
    cross sectional view, 34f
    cultivation of E. coli, 35
    electron micrograph of densely
      packed N. mediterranei, 41f
    schematic diagram of experimental
      setup, 36f
    immobilization of enzymes, 59
  membrane, advantages and
    disadvantages, 32
Hydrocortisone, transformation to
  prednisolone, aqueous two-phase
  system, 83
Hydrogel, effect of bioproduct
  diffusivity on ligand
  consumption, 164f
Hydrogel beads
  bioproduct separation, immobilized
    affinity adsorbents, 153
  simulation studies, small affinity
    adsorbent
    immobilization, 158-164

Hydrolysis
 cell wall, enzymatic lysis and
  disruption of yeast cells, 18
 glucan and soluble protein,
  enzymatic lysis and disruption
  of yeast cells, 16
 triglyceride, membrane reactor,
  enzymatic, 58
Hydrophobic chromatography, polymer
 removal, protein purification, 89

I

Immobilization
 aerobic whole-cell, dual
  hollow-fiber bioreactor, 33
 effect on antibody activity, 209
 enzymes in membranes, 59
 materials and methods, EPR
  spectroscopy, 210
 monoclonal antibodies, linkage
  through protein A, 214
 simulation studies, small affinity
  adsorbents in hydrogel
  beads, 158-164
Immobilized affinity adsorbents
 mathematical modeling of bioproduct
  adsorption, 153
 schematic diagram, 156f
Immobilized allosteric enzymes, gelled
 membrane formation, 61
Immobilized monoclonal antibodies
 EPR studies of structure and
  function, 208
 simulated and experimental spectra,
  EPR, 212
Immobilized systems
 alcoholic fermentations compared to
  aqueous two-phase systems, 80
 fine chemicals production compared
  with aqueous two-phase
  systems, 82
Immunoglobulin
 description, 209
 protein recovery by salt
  partition, 107
Immunoglobulin G
 concentration in salt phase as
  function of salt concentration,
  PEG-water solution, 100f
 partition coefficient as a function
  of pH, PEG-water solution, 101f
 partition coefficient as function of
  salt concentration, PEG-water
  solution, 100f
Inhibition, fermentation processes,
 extractive bioconversions in
 aqueous two-phase systems,
 product, 79-85

Interferon, RIEF, 191
Ion exchange chromatography, polymer
 removal, protein purification, 89
Ion exchange resins as chromatographic
 supports, scale-up of
 chromatographic processes, 124
Isoelectric focusing (IEF)
 apparatus, recycling, 187
 electrophoresis scale-up
  considerations, 139
 scale-up, discussion, 185
Isolation of enzymes in aqueous
 two-phase systems, 85t

J

Joule heating
 classic thin-film CFE, 170
 viscosity, and buoyancy,
  electrophoresis scale-up
  considerations, 138

L

Large-scale gel chromatography,
 purification of protein
 products, 193
Large-scale protein fractionation,
 RIEF, 185
Linear equilibrium, size exclusion
 chromatography, 132
Linkage through immobilized protein,
 immobilized monoclonal
 antibodies, 214
Liquid chromatography, process
 considerations for scale-up, 122
Liquid emulsion membrane (LEM)
 amino acid recovery
  discussion, 71-75
  process diagram, 76f
 applications, 70
 biochemical separations, 67
 chloride countertransport
  system, 73f
 concept, 67
 effect of agitation speed, 73f
 system, 69f
 versatility, 75
Loading, gel chromatography design
 model, 198
Lysate, amino acid, membrane recovery
 process, 56
Lysing yeast cell, schematic, 12f
Lysis
 sequence of cell, enzymatic lysis
  and disruption of yeast cell, 11
 yeast, process conditions for enzyme
  release simulation, 26

INDEX

Lytic enzyme system
  enzymatic lysis and disruption of
    yeast cell, 11
  protein release from yeast cells, 10
  yeast lysis, 16

M

Macromolecules, production from
  aqueous two-phase systems, 84
Mass transfer expressions, size
  exclusion chromatography, 131t
Mechanical stability, LEMs, 68
Mechanically based protein release
  methods, undesirable properties, 2
Membrane
  asymmetric enzyme capillary, phase
    inversion method, 63f
  enzyme, research and development, 59
  liquid emulsion, amino acid
    recovery, 71-75
Membrane bioreactors
  amino acid from lysate, 56
  bioactive compounds from
    fermentation broths, 52
  enzyme, 58
  growth of rumen bacteria
    glucose, 45t
    sugar beet pulp, 46t
  hollow fiber, advantages and
    disadvantages, 32
  SCP and methane, rumen
    fermentation, 47f
  simultaneous production of SCP and
    methane, 43
Membrane distillation, summary,
  biotechnology, 55t
Membrane formation, S. solfataricus,
  gelled enzyme, 61-65
Membrane formulation, LEM-mediated
  amino acid recovery, typical, 72
Membrane processes
  bioactive compounds from
    fermentation broths, 52
  downstream, 57f
  summary, biotechnology, 55t
Membrane swell, LEMs, 70
Membrane technology, integration with
  aqueous two-phase systems, 82

Methane
  membrane bioreactor production, 43
  SCP and, rumen fermentation
    production in membrane
    bioreactor, 47f
Microbial cells, subcellular location
  of enzyme activities, 10
Microfiltration, summary,
  biotechnology, 55t

Mitochondria, release of cytosol and,
  enzymatic lysis and disruption of
  yeast cells, 18
Model
  convective dispersion, RCFE, 174
  enzymatic lysis and disruption of
    yeast cells
    applications, 24
    background, 11
    discussion, 9
    soluble protein, peptides, and
      carbohydrates, 21-24
    yeast mass, 21-24
  particle size distribution, protein
    precipitation, 112-116
  precipitation phenomena, protein
    recovery, 109
  yeast lysis
    reaction pathways for
      structured, 17f
    simple and structured, 13
    structured, 19
    variables and parameters, 15t
Monoclonal antibodies
  immobilized, EPR studies of
    structure and function, 208
  materials and methods, EPR
    spectroscopy, 210,214
Multiphase aqueous systems, phase
  partitioning, 95

N

Nocardia mediterranei
  electron micrograph, dual
    hollow-fiber bioreactor, 41f
  immobilization, dual hollow-fiber
    bioreactor, 32
  production of rifamycin B, dual
    hollow-fiber bioreactor, 37
Nucleic acid
  partitioning, multiphase aqueous
    systems, 96
  protein recovery by salt
    partitioning, 107

O

Oligosaccharides, ion exchange
  separation, 125f
Organic solvent, effect, enzymatic
  synthesis of fine chemicals, 83
Osmolality of cell lysate, effective,
  enzymatic lysis and disruption of
  yeast cells, 16
Osmotic swell, LEMs, 70

P

Particle formation, primary, protein
   precipitation, 110
Particle size distribution, protein
   precipitation, model, 112-116
Partition coefficient, PEG-water
   solution, immunoglobulin as a function
   of pH, 101f
Partitioned phases, applications, 96
Partitioning
   affinity, aqueous two-phase
      systems, 86t
   affinity sorbent, two-phase
      system, 88f
   plasma proteins, multiphase aqueous
      systems, 96
   proteins, aqueous two-phase systems,
      affinity, 85-87
Peclet numbers, RCFE, performance
   evaluation, 178-181
Peptides and carbohydrates, enzymatic
   lysis and disruption of yeast
   cells, 21-24
Permeabilization, chemical, protein,
   DNA, and RNA release from
   E. coli, 2-4
pH
   effect on salt partition recovery of
      proteins, 103
   effluent, dual hollow-fiber
      bioreactor, glucose, 36f
   gradients, RIEF, 189-191
Phase inversion, asymmetric enzyme
   capillary membranes, 63f
Phase partitioning, multiphase aqueous
   systems, 95
Phenylalanine, downstream processing
   from fermentation broth,
   LEM-mediated separations, 71
Phenylalanine-chloride system,
   mechanism for facilitated
   transport, 69f
Plasma proteins, partitioning,
   multiphase aqueous systems, 96
Plate height and resolution concepts
   in electrophoresis,
   scale-up, 141-143
Poly(ethylene glycol) (PEG)-dextran
   extractive fermentation, two-phase
      system, 80
   two-phase systems in
      biotechnology, 78
Poly(ethylene glycol) (PEG)-water,
   recovery of proteins by salt
   partition, 93,104-107
Polyclonal antibody fractionation,
   RIEF, 189,190f
Polymers, removal from purified
   protein, aqueous two-phase
   systems, 87

Population balances, modeling, protein
   precipitation, 113
Precipitate behavior, models, protein
   precipitation, 116
Precipitation
   model
      particle size distribution,
         protein, 112-116
      protein recovery, 109
Prednisolone, aqueous two-phase
   system, transformation of
   hydrocortisone, 83
Product inhibition, fermentation
   processes, extractive
   bioconversions in aqueous
   two-phase systems, 79-85
Production
   bulk chemicals, two-phase
      systems, 80
   fine chemicals, aqueous two-phase
      systems, 82
   macromolecules, aqueous two-phase
      systems, 84
   protective antigens, aqueous
      two-phase systems, 84
   rifamycin B by N. mediterranei, dual
      hollow-fiber bioreactor, 37
Production-scale electrophoresis
   system, center for separation
   science, 191
Production-type gel chromatography
   system, sequence of
   operations, 197f
Productivity, gel chromatography
   design model, 210,202f
Protein
   anaerobic SCP, membrane reactor
      production, 43
   aqueous two-phase systems, removal
      of polymers, 87
   large-scale gel chromatography
      fractionation, 193
   partitioning of plasma, multiphase
      aqueous systems, 96
   recombinant, problems in
      production, 9
   release, E. coli, 4
   RIEF, 191
Protein fractionation
   RIEF, large-scale, 185
   scale-up methods, gel
      chromatography, 195f
Protein hydrolysis, enzymatic lysis
   and disruption of yeast cells, 16
Protein precipitation
   discussion, 110-112
   particle size distribution, 112-116
   precipitate behavior, 109,116
Protein purification
   inherent problems, gel
      chromatography, 194
   partitioning in aqueous two-phase
      systems, 85-87

Protein recovery
  effect of pH and salt concentration
    on salt partitioning, 103
  salt partitioning
    effect of salt
      concentration, 97,98
    PEG-water solution, 93,104-107
Protein release
  chemically permeabilized
    E. coli, 2-7
  Triton and guanidine HCl treatment
    of E. coli, 6f
  undesirable properties,
    mechanical, 2
Protein separation
  chromatographic, calculated elution
    profile, 137f
  electrochromatographic, calculated
    elution profile, 145
Proteolysis, after mechanical
  disruption, 10
Purification of proteins by
  partitioning in aqueous two-phase
  systems, 85-87

R

Reaction pathways for structured
  model, yeast lysis, 17f
Recombinant proteins
  problem in production, 9
  recovery from E. coli, 2
Recovery and purification in
  biotechnology, aqueous two-phase
  systems, 78-90
Recycle continuous-flow
  electrophoresis (RCFE)
  discussion, 171-173
  model, 173-178
  performance evaluation, elevated
    Peclet numbers, 178-181
  regenerators, schematic, 182f
  schematic, 172f
Recycling isoelectric focusing (RIEF)
  illustrative fractionation
    results, 187-189
  polyclonal rabbit antibodies, 190f
  schematic presentation, 188f
Regenerators, RCFE, 181
Reverse osmosis
  bioactive compounds from
    fermentation broths, 52
  summary, biotechnology, 55t
Rifamycin B
  continuous production, dual
    hollow-fiber bioreactor, 40f
  production by N. mediterranei, dual
    hollow-fiber bioreactor, 37

RNA release
  E. coli treated with guanidine HCl
    and Triton, 5f
  mechanically disrupted E. coli, 4
Rumen bacteria, growth on glucose,
  membrane reactor, 45t
Rumen fermentation, membrane
  bioreactor, SCP and methane
  production, 47f
Rumen microorganisms, anaerobic SCP
  production, 43

S

Salt concentration, effect, salt
  partitioning recovery of
  proteins, 98,103
Salt partition
  applications, protein
    recovery, 104-107
  effect of pH and salt concentration,
    protein recovery, 103
  methodology, protein recovery, 97
Scale-up considerations
  chromatographic processes, 126-129
  cost analysis, gel
    chromatography, 204
  gel chromatography, protein
    fractionation, 195f
  IEF, 185
Selectivity equation, gel
  chromatography design
  model, 198,200f
Separations
  product from key contaminant, gel
    chromatography, 195f
  using LEMs, 68
Separations science, center for,
  production-scale electrophoresis
  system, 191
Single-cell protein (SCP)
  anaerobic, membrane reactor, 43,44
  methane, rumen fermentation in
    membrane reactor, 47f
Single-component diffusion and
  binding, immobilization in
  hydrogel, 158-163
Single-pass continuous flow
  electrophoresis, model
  comparison, 175f
Size-exclusion chromatography
  initial and boundary
    conditions, 130t
  mass transfer and equilibrium
    expressions, 131t
  scale-up, theoretical
    considerations, 129-135
Solfolobus solfataricus, gelled enzyme
  membrane formation, 61-65

Soluble products, model, enzymatic
  lysis and disruption of yeast
  cells, 18
Soluble protein
  hydrolysis, enzymatic lysis and
    disruption of yeast cells, 16
  peptides, and carbohydrates,
    enzymatic lysis and disruption
    of yeast cells, model
    simulations, 21-24
Solvents, production, extractive
  fermentation in aqueous two-phase
  systems, 80
Spin label, EPR sensitivity,
  immobilized monoclonal
  antibodies, 212
Spin label motion, monoclonal
  antibodies, 210
Sugar, ion exchange separation, 125f
Sugar beet pulp, membrane reactor,
  growth of rumen bacteria, 46t
Sulfuric acid, ion exclusion
  separation, 127f
Swelling, LEMs, 70

T

Thin-film continuous-flow
  electrophoresis, general
  discussion, 170
Toxicity, macromolecules on bacterial
  cells, 84
Transport, LEMs, 68
Triglycerides, degree of conversion
  with time, membrane processes, 60f
Triton, cellular protein release from
  E. coli, 4
Two-component diffusion and binding,
  immobilization in hydrogel, 163
Two-phase system
  affinity sorbent, partitioning, 88f
  aqueous
    biocatalytic reaction, 79
    bulk chemical production, 82t
    examples of affinity
      partitioning, 86t
    extractive bioconversions, product
      inhibition in fermentation
      processes, 79-85
    fine chemicals production, 82
    isolation of enzymes, 85t
    macromolecules production, 84
    principle for extractive
      bioconversion, 81f
    purification of proteins by
      partitioning, 85-87
    recovery and purification in
      biotechnology, overview, 78-90
Two-phase systems, purification and
  recovery in biotechnology, 78

U

Ultrafiltration (UF)
  bioactive compounds from
    fermentation broths, 52
  summary, biotechnology, 55t

V

Viscosity, joule heating, and
  buoyancy, electrophoresis scale-up
  considerations, 138
Volumetric productivities, industrial
  and large-scale gel chromatography
  protein fractionations, 195t

W

Wall hydrolysis equations, enzymatic
  lysis and disruption of yeast
  cells, 18
Water-soluble polymers, two-phase
  systems in biotechnology, 78

Y

Yeast cells
  applications, model of enzymatic
    lysis and disruption, 24
  models, enzymatic lysis and
    disruption, 9
  release of cytosol and
    mitochrondria, 18
  schematic of lysing, 12f
  soluble products, model, enzymatic
    lysis and disruption, 18
  structure model, enzymatic lysis and
    disruption, 11
Yeast lysis
  enzyme recovery from subcellular
    structures, 27f
  lytic system, enzymes, 16
  model, process conditions for enzyme
    release simulation, 26
  release of site-specific enzymes,
    simulation, 25f
  simple model, 13,17f
  structured model, 13
  variables and parameters, model, 15t
Yeast mass, enzymatic lysis and
  disruption of yeast cells, model
  simulations, 21-24
Yeast wall, double-layered
  structure, 12f
Yield and purity, gel chromatography
  design model, 198

Production by Joan C. Cook
Indexing by Susan Robinson
Jacket design by Pamela Lewis

Elements typeset by Hot Type Ltd., Washington, DC
Printed and bound by Maple Press Co., York, PA

RECENT ACS BOOKS

"Chemistry and Function of Pectins"
Edited by Marshall Fishman and Joseph Jen
ACS Symposium Series 310; 286 pp; ISBN 0-8412-0974-X

"Fundamentals and Applications of Chemical Sensors"
Edited by Dennis Schuetzle and Robert Hammerle
ACS Symposium Series 309; 398 pp; ISBN 0-8412-0973-1

"Polymeric Reagents and Catalysts"
Edited by Warren T. Ford
ACS Symposium Series 308; 296 pp; ISBN 0-8412-0972-3

"Excited States and Reactive Intermediates:
Photochemistry, Photophysics, and Electrochemistry"
Edited by A. B. P. Lever
ACS Symposium Series 307; 288 pp; ISBN 0-8412-0971-5

"Artificial Intelligence Applications in Chemistry"
Edited by Bruce A. Hohne and Thomas Pierce
ACS Symposium Series 306; 408 pp; ISBN 0-8412-0966-9

"Organic Marine Geochemistry"
Edited by Mary L. Sohn
ACS Symposium Series 305; 440 pp; ISBN 0-8412-0965-0

"Fungicide Chemistry: Advances and Practical
Applications"
Edited by Maurice B. Green and Douglas A. Spilker
ACS Symposium Series 304; 184 pp; ISBN 0-8412-0963-4

"Petroleum-Derived Carbons"
Edited by John D. Bacha, John W. Newman and
J. L. White
ACS Symposium Series 303; 416 pp; ISBN 0-8412-0964-2

"Coulombic Interactions in Macromolecular Systems"
Edited by Adi Eisenberg and Fred E. Bailey
ACS Symposium Series 302; 272 pp; ISBN 0-8412-0960-X

"Historic Textile and Paper Materials: Conservation
and Characterization"
Edited by Howard L. Needles and S. Haig Zeronian
Advances in Chemistry Series 212; 464 pp; ISBN 0-8412-0900-6

"Multicomponent Polymer Materials"
Edited by D. R. Paul and L. H. Sperling
Advances in Chemistry Series 211; 354 pp; ISBN 0-8412-0899-9

For further information contact:
American Chemical Society, Sales Office
1155 16th Street NW, Washington, DC 20036
Telephone 800-424-6747